The JOY of Sciencing:

A Hands-on Approach to Developing Science Literacy and Teen Leadership through Cross-Age Teaching and Community Action

The 4-H SERIES Project in Action

Edited
by Richard Ponzio
and Charles Fisher

The JOY of Sciencing:

A Hands-on Approach to Developing Science Literacy and Teen Leadership through Cross-Age Teaching and Community Action

The 4-H SERIES Project in Action

Edited by Richard Ponzio and Charles Fisher

SERIES is an acronym for Science Experiences and Resources for Informal Education Settings, a 4-H program nationally headquartered at the University of California, Davis, and active at sites across the United States.

The 4-H Science And Youth (SAY) Project is funded by a grant from the National Science Foundation. This material is based upon work supported by the National Science Foundation under Grant #ESI-9254695. The government has certain rights in this material. Any opinions, findings, and conclusions or recommendations expressed in this material are those of the authors and do not necessarily reflect the views of the National Science Foundation.

Published for the National 4-H SERIES Office, Department of Human and Community Development, University of California, Davis, California 95616, U.S.A.
by Caddo Gap Press, 3145 Geary Boulevard, Suite 275, San Francisco, California 94118 U.S.A., Alan H. Jones, Publisher

ISBN 1-880192-24-1
List Price: $19.95

Cover photograph shows the 4-H SERIES "ADOPT-A-BEACH RESEARCH" project crew at Christies' Beach, Santa Cruz Island, California, August, 1994. All photographs provided by and property of The 4-H SERIES Program.

For information about the 4-H SERIES or Science And Youth Projects contact the 4-H SERIES Project, Department of Human and Community Development, University of California, Davis, California 95616, U.S.A.; telephone 530/752-8824. Additional copies of this publication may be ordered from the 4-H SERIES Project or from Caddo Gap Press, 3145 Geary Boulevard, Suite 275, San Francisco, California 94118, U.S.A.; telephone 415/392-1911; fax 415/392-8748; e-mail caddogap@aol.com

Table of Contents

Acknowledgement

The manuscripts that make up *The Joy of Sciencing* would never have been written—the events and outcomes depicted in these pages would never have happened—without the generous support of the National Science Foundation which believed in the power of teenage volunteers as guides and teachers of children. The National Science Foundation's investment in tomorrow's leaders bore fruit in rural, suburban, and urban communities from California to Maine. Tens of thousands of teenage volunteers and their adult coaches were encouraged to pursue the joy of learning science by exploring natural phenomenon. The adult coaches supported and mentored the teenage volunteers who were eager to share their newly discovered knowledge and investigative skills with hundreds of thousands of children.

Hundreds of school and community-based programs across the nation would never have had access to the 4-H SERIES project without the support of the University of California's 4-H Youth Development Program, 4-H colleagues, and classroom teachers who participated in the development of the materials and instructional strategies used in the Science And Youth Project. The National 4-H Youth Development Program helped shepherd the program across the nation, and contributed to the development of 4-H Regional Leadership Centers for Science And Youth at land grant universities in California, Georgia, Kentucky, Missouri, and New York.

Teen volunteers and children would never have experienced the joy of sciencing or using their science skills to contribute to the spirit of their community without the countless hours unselfishly given by adult volunteers, often teachers or family members, who served as coaches, guides, and friends. Teen leaders found joy and satisfaction in working with children who spoke different languages, came from different cultures, and lived in different neighborhoods. Side-by-side they discovered science in the world around them and applied science thinking processes and insights to their everyday lives. Together—learning science—they found a common ground—making their community a better place to live. We hope this book inspires you, too—to find joy in sciencing.

Sheryl — Thanks for the exceptional talent and insight you bring to the WELS team. You're special! Cheers, Ed

The Editors

Richard Ponzio is the Science and Technology Specialist for the California 4-H Youth Development Program and a faculty member in the Department of Human and Community Development at the University of California at Davis. His areas of investigation are the development of science literacy programs and assessment of science education in non-school settings as well as the involvement of families in community-based education.

Charles Fisher is a research scientist in the Educational Studies Program in the School of Education at the University of Michigan. His work examines how learners carry out literacy and science tasks in both formal and informal learning environments.

Ponzio and Fisher have sailed in wooden boats together for more than fifteen years and during that time they have come to certain realizations about education, joy, and the meaning of life. They believe learning is like sailing in at least three ways: it's best when done outdoors; it's wonderful when done in the company of friends; and you can never do too much of it. In putting this book together they also came to realize the only sailor to have his work done by Friday was Robinson Crusoe.

Contributing Authors

David C. Berliner, Regents' Professor of Education, College of Education, Arizona State University, Tempe, Arizona

Dale Cox, Regional 4-H Youth Specialist, University Outreach and Extension, University of Missouri, Bolivar, Missouri

Rachel Davis, County Extension Agent for 4-H/Youth Development, University of Kentucky, Lexington, Kentucky

Charles Fisher, Senior Research Scientist, School of Education, University of Michigan, Ann Arbor, Michigan

Valerie A. (Pankow) Joyner, Elementary School Teacher, Petaluma City Schools, Petaluma, California

Janice H. Poda, Director, South Carolina Center for Teacher Recruitment, WU Station, Rock Hill, South Carolina

Richard Ponzio, 4-H Specialist and Faculty Member, Department of Human and Community Development, University of California, Davis, California

Introduction

By David C. Berliner

I have come to believe that you cannot go far astray when baking cookies if you use lots of butter, chocolate, and sugar. Whatever comes from a recipe with ingredients as sumptuous as these is likely to be pretty darn good! So it is with the components of the various instructional programs described in this volume. The instructional ingredients in these programs are often among the most desirable we can find—like butter and chocolate. And just as in baking, if the right instructional ingredients are blended well they rarely fail to yield end products that are remarkably good.

What are the instructional ingredients in the programs described in this volume? They include:

♦ *Project based instruction.* Projects of duration, depth, and complexity elicit greater motivation from students and seem to develop greater levels of understanding than do conventional methods of instruction.

♦ *Relevance to the real world.* The kinds of authenticity described are often lacking in ordinary classroom instruction.

♦ *A high probability of obtaining rewards by the students from people they admire in their own communities.* This occurs because the students are able to find, frame, and address problems that need attention in the communities in which the students live.

♦ *Hands-on learning.* This quality is too often missing from conventional instruction.

◆ *Intrinsically interesting curricula and tasks, learned in informal settings.* These characteristics enhance motivation. The fact that the curricula is all about science adds to the potential that the skills learned in the informal settings will carry over into school settings and other real world settings that require small group interactions for planning, analyzing, and solving genuine problems.

◆ *Instructors who are teens.* The cross-age teaching that is characteristic of these programs promotes two desirable outcomes. First, younger children find the teens easier to relate to and emulate than they would an adult teacher. This promotes learning and identification with the values expressed by schools through the teen instructors. Second, the teens get real world teaching experience, which promotes thoughtfulness and responsibility on their part. We all have come to learn that teens need to have access to groups that are engaged in prosocial activities. If not, they sometimes end up in groups engaged in antisocial activities. In these programs teens who volunteer have something satisfying to do.

These ingredients—project method, relevance and authenticity, high probability of reward and feeling of usefulness, tasks of intrinsic interest, informal settings for learning, cross-age teaching—are all of the highest quality. They are the butter, chocolate, and sugar of baking. And to finish with the metaphor, the recipes have been tested and work. Moreover, the cost of the ingredients is minimal—these programs are inexpensive to run.

I am impressed also that these programs have been tinkered with over a number of years. They have been field-tested and redesigned so that they have a high probability of success. Teenagers can and do teach effectively with these materials, a point made persuasively by the participants, focus groups, supervisors, parents, and others in the evaluative data that have been collected. Teens typically come away saying that they are better for the experience—more self confident, more responsible, better communicators, more understanding and respectful of their teachers, and even more likely to go into teaching.

The youngsters that are taught are also enriched and empowered by their participation. Although the data indicate that a number of improvements can be made, it is hard to be anything but positive about the programs described. With the current versions, large numbers of youthful participants, male and female, minority and majority, are learning science *and* playing, doing useful things *and* having a good time, developing values that are highly desirable *and* enjoying the process, serving their communities *and* having gratifying social interactions. I expect that readers of this book will join me in celebrating the success of these innovative educational programs.

It should also be noted that these are the kinds of programs that are strongly advocated by Robert Reich, former U.S. Secretary of Labor. He cogently argues that the educational experiences of youth must foster problem identification as well as problem solving—skills that are acquired though participation in these programs. Reich has also written that the educational experiences for a smooth transition from school to work are experiences that teach students:

◆ responsibility;
◆ the skills for collegial work;
◆ the need for planning;
◆ the ability to manage resources; and
◆ the processes of assessment.

These are among the skills learned in the programs described in this book.

Thus, it is obvious that what is learned in these programs is often quite different from what is learned in traditional school-based science. By learning to apply and reflect on science in the context of real world problem solving, youth learn that scientific thought and activities are not merely arcane school subjects but powerful ways of organizing the experience of our daily lives. This high potential for transfer of scientific ways of thinking to real life situations could have two important consequences: first, it may provide the best chance for U.S. industry to maintain its competitive edge in the global economy; and second, it may increase the likelihood that the citizenry of the U.S. will solve more of its social and environmental problems in reasonable and responsible ways.

I hope that readers of this volume will join me in recommending the dissemination of both the materials and the processes of learning that are described in the following chapters. These programs are a very welcome addition to the educational armamentarium of our nation.

Teen leader demonstrates seismic wave movement during 4-H SERIES session at the California Academy of Science in San Francisco.

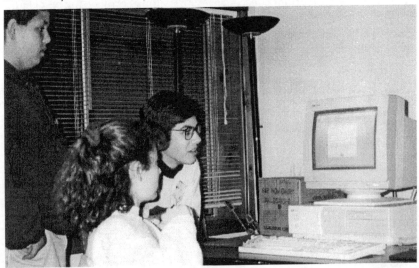

At the Walker Creek Environmental Ranch in Petaluma, California, 4-H SERIES teens communicate with other teens across the country via a computer link set up by Richard Mahacek, 4-H youth development advisor in Merced County, using computers and software donated by the Hewlett-Packard Company.

Chapter 1

The 4-H SERIES Program:
An Overview

By Richard Ponzio, Charles Fisher, & Valerie (Pankow) Joyner

In 1988, a team of scientists and science educators began to design, develop, and test an innovative strategy for teaching science concepts and processes. 4-H SERIES was designed to bridge the gap between the science classroom and everyday life. 4-H SERIES brought together several proven educational practices in a unique juxtoposition. It combined cross-age tutoring with adult mentoring, hands-on instruction and community service with the National Science Education Standards. The strategy included the development of six multiple-lesson curriculum units (later expanded to eight units). Each unit focused on a set of highly-valued science topics while engaging learners in performing seven key science processes. The hands-on, highly interactive science activities were designed to promote scientific inquiry among children in the 9- to 12-year-old age range. The activities, led by specially-trained teenagers, were presented in a variety of non-school or informal education settings. As part of the design, children and their teen leaders identified community issues related to the content of the curriculum unit they were working with and subsequently developed and implemented a community service project to address the identified issue.

Taken as a whole, the curriculum units, the inquiry-oriented pedagogy, the cross-age teaching, the community service component, and other aspects of the design are formally known as Science Experiences and Resources for Informal Education Settings or SERIES. Initial development was funded by the National Science Foundation, and the project was completed under the auspices of the 4-H Youth Development Program of the University of California. Beginning primarily in 4-H clubs, camps, fairs, and teens' backyards, the program spread quickly in both informal and formal educational settings. By 1996, there were 4-H SERIES programs in 47 states, Puerto Rico, and Guam, involving thousands of teens and tens of thousands of younger learners in hands-on science activities in remarkably diverse settings.

The purpose of this monograph is to describe, at least in part, some examples of how, where, when, and with whom, the 4-H SERIES program has been implemented. This introductory chapter provides background information on the 4-H SERIES program, an early assessment of its impact on teens and younger youths, a description of how and why the cases included in subsequent chapters were selected, and an outline of relationships among and between the cases. Following this introduction, chapters two through six each represent a brief case study of a specific implementation of 4-H SERIES. A concluding seventh chapter discusses the value of such informal science education programs.

Emergence of the 4-H SERIES Program

Beginning in the early 1980s, there was renewed interest in increasing the quality and quantity of science experiences available to American youth. This new wave of enthusiasm for science and science literacy was driven by a variety of potent factors. In the educational arena, inquiry methods of science teaching were regaining adherents, social learning theories were increasingly favored, educators began to embed science instruction within authentic social contexts, and traditional categories of scientific knowledge were breaking down. Changes in science education were bolstered by changing expectations in jobs and patterns of employment. Teenagers were no longer preparing for a single career, globalization was affecting job markets, emphasis on workgroup dynamics called for leadership skills, and new technologies required new cognitive and communication skills.

By the mid 1980s, political and educational leaders had articulated a set of national education goals calling for, among other things, dramatic improvement in the science literacy of children in kindergarten through high school. Within this context, a team of educators at the University of California proposed the 4-H SERIES Project. With partial funding from the National Science Foundation in 1988, SERIES' innovative approach to learning science was developed and piloted in California. In 1991, a second grant from the NSF funded national dissemination of 4-H SERIES through the creation of Regional Leadership Centers in New York, Georgia, Missouri, and California.

At the heart of 4-H SERIES there is a belief that learning science is not merely learning the facts of science, or learning more information about science. Rather, what is valued is *doing* science; applying and practicing the strategies of scientific thinking, gathering evidence and using decision-making skills—in a word—"sciencing." 4-H SERIES is a community-based informal science program that engages 9 to 12 year-old youngsters in these "sciencing" activities, thereby bridging the gap between science and everyday life. Designed for participants of different ages and abilities, it allows children to learn science from each other.

Specially trained teenage volunteer leaders interact with the younger children as they inquire into natural phenomena, record data, make inferences, and so on. Teen leaders are able to work effectively with youngsters because their age differences are small and, since teen leaders usually work in teams, individualized attention is almost always available.

Another basic component of the 4-H SERIES program is the use of teens as teachers and adult volunteers as coaches. The 4-H SERIES Project staff recruits and trains adult volunteers who, in turn, coach teen leaders as they lead actual science experiences for the 9-to-12-year-old participants. Volunteers are often practicing scientists and engineers who devote their time to working with 4-H SERIES youth. 4-H SERIES has benefited from relationships with several science-oriented professional organizations, especially the Society of Women Engineers, that have made valuable contributions to both the 4-H SERIES program and thousands of participating youths.

The 4-H SERIES curricula include eight topic-centered units covering a wide variety of scientific phenomena and principles. Each unit of the curricula is built on a model of instructional design known as the learning cycle (Karplus, Lawson, Wollman, Appel, Bernoff, Howe, Rusch, and Sullivan, 1980; Karplus and Thier, 1967; Lawson, Abraham, and Renner, 1989). This model is widely used in contemporary science education and has shown consistent learning gains in a variety of settings (Guzzetti, Snyder, and Glass, 1992). The 4-H SERIES curricula also emphasize use of the scientific thinking processes that are found in most state department of educaton guidelines and are included in the vast majority of science texts in the United States. These processes—observing, communicating, comparing, organizing, relating, inferring, and applying—are used by participants in each of the activities. Titles of the units and the general topics that they deal with are presented in Table 1.

Groups of 9-to-12-year-olds, led by older teens, work together in a sequence of about six sessions per curriculum unit. Each session is approximately one to two hours long and may take place practically anywhere, such as at a 4-H club meeting, an after-school club, a youth organization activity, or as part of a parks and recreation program. The consumable materials used in SERIES' activities, such as baking soda, vinegar, and even backyard snails, are inexpensive and readily obtained, making it easy for participants to continue science investigations on their own. Most units are available in Spanish as well as English. As part of the group's work, the teen leaders and the younger youth design and carry out a community service project related

Table 1

4-H SERIES Curriculum Units

Curriculum Title	Topic Area
Beyond Duck, Cover, and Hold	Earthquake and disaster preparedness.
Chemicals Are Us	What chemicals and chemical reactions are and how to use chemicals responsibly.
Recycle, Reuse, Reduce	Natural cycles, ways to recycle and reuse materials, reduce amounts of materials discarded.
What's Bugging You?	Pests and responsible pest management.
Snailing and Sciencing	Learning to use seven scientific thinking processes.
Oak Woodland Habitats	Environmental interdependence.
It Came from Planted Earth	Agriculture, plant food, and fiber uses.
From Ridges to Rivers	Watershed explorations.

to the science topic being studied and appropriate to local neighborhood needs, such as a creek clean-up, disaster preparedness kits, planting a community garden, etc.

While the activities in the units are enjoyable and interesting, the cornerstone of the curricula is the manner in which the activities are carried out. The activities are typically conducted in groups with extensive interaction among participants and teen leaders. 4-H SERIES activities are characterized by a kind of playfulness that invites exploration, experimentation, and learning.

4-H SERIES was designed to get groups of people together—teenagers, younger children, adult volunteers, and working scientists—who do not typically work with one another, to talk about and do science in a variety of community settings. Most often teenagers are in charge and young people have an opportunity to make authentic contributions to their communities. 4-H SERIES encourages the kind of dynamic group problem solving that goes on in the work place rather than having participants complete exercises and answer questions in a formal environment controlled by adults.

For the teen leaders, there are several benefits. They receive training in science content and gain leadership experience by taking responsibility for the activities with younger participants. These responsibilities provide authentic motivation for doing the job well and genuine satisfaction when the work is completed successfully. Since teens typically work in teams while planning and implementing the program, a large part of 4-H SERIES constitutes leadership development. Teens have the opportunity to work with practicing scientists and engineers, often visiting their places of work and getting a close-up review of what a career in science is really like.

A professional engineer from the Society of Women Engineers assists girls at Charles Armstrong Middle School in Belmont, California, discover what a snail will and will not eat.

4-H SERIES Outreach

4-H SERIES activities take place in a wide variety of settings including 4-H clubs, Boys and Girls Clubs, after school programs, county and local fairs, summer camps, urban housing projects, community science programs, migrant education programs, and museums and schools, among others. The Association of Science and Technology Centers' Youth ALIVE! Program uses SERIES materials as an integral feature of its training program for teen leaders. The National Society of Women Engineers (SWE) features 4-H SERIES as a major program for their members to help improve the scientific literacy of America's youth, and to encourage women and minorities to pursue careers in science and mathematics. The National Red Cross uses 4-H SERIES materials and the teen leadership model in its Youth Emergency Service program.

Following rapid development in California and use across the nation, 4-H SERIES has continued to expand on several dimensions. A number of derivative programs have arisen as new groups of possible participants have come in contact with the program. Thumbnail descriptions of two of these programs will indicate 4-H SERIES appeal to, in one case, formal education systems and, in another, to school-age child care providers.

The Science and Youth (SAY) Project

A number of high schools throughout the United States have special magnet programs designed around vocations, performing arts, subject matter specialties, and so on. These programs operate like schools within a school. Approximately 20 high schools have programs of this type that are designed for students who are interested in exploring teaching as a career. These programs, called teaching academies, provide a curriculum strand that blends academic content with learning about pedagogy. Thirteen of these teaching academies participate in the SAY project by using 4-H SERIES curriculum and training methods to structure early science teaching experiences for their students. In the SAY project, high school students use 4-H SERIES lessons to teach classes of elementary students in local classrooms. In this way, student-centered and problem-centered science activities are introduced to substantial numbers of elementary and secondary schools. By participating in the SAY project, hundreds of prospective science teachers get early experiences teaching science using constructivist teaching and learning methodologies while working with real students.

In most cases, the teaching academy is housed at one high school in the district and draws students from throughout the district. A unique exception to this pattern occurs in South Carolina where there are 133 high schools and each has most of the components of a teaching academy. In addition to coursework, participants in the teaching academy programs typically engage in tutoring and provide other educational experiences for younger children. Each academy usually has some kind of partnership with one or more local teacher training institutions. Two of the cases presented later in this monograph focus on SAY project sites.

The Youth Experiences in Science (YES) Project

The YES project is producing, testing, and disseminating an informal science education curriculum expressly designed for 5-to-9-year-olds in school-age child care (SACC) settings. Six thematic units have been developed to date, and a seventh unit is currently being developed. YES plans to have all seven units available in English and Spanish versions. The units incorporate the basic design process and instructional model developed in the earlier 4-H SERIES Project. Each unit consists of a sequence of engaging hands-on activities performed with inexpensive materials and incorporating community service applications. During the three years of YES's NSF support, approximately 5,500 teenage volunteers will deliver the program to 20,000 children through a network of existing SACC sites operated by 4-H and other community groups throughout California. Capitalizing on SACC settings, the program is collaborating with schools and reaching out to the parents of participating children through "Science Family Backpacks" and science family activity nights.

In addition to formative and summative evaluation, data is being gathered on parent involvement and the amount, nature, and durability of positive influences on participants' school science success. Evaluation of teen-leader effectiveness in conducting school-age child care science programs has shown that teens bring a unique dimension to the programs, benefiting the young children, the teens,

Friendships are formed in 4-H SERIES programs between teen leaders and younger children.

and also the regular site staff. Findings from these studies are currently in press, and a national YES model is under consideration.

Summary

Since 1988, 4-H SERIES has grown from a local project in California to a national program for developing science literacy and leadership skills by engaging youths in hands-on science activities that relate to their everyday lives and the communities in which they reside. 4-H SERIES has made its largest impact in informal education settings including 4-H venues, museums, and a wide variety of youth organizations. 4-H SERIES has also been widely used as a key component of science programs in formal education settings, teacher education, and school-age child care programs, among others. In late 1996, 4-H SERIES was being used in 47 states, Puerto Rico, and Guam. It has proven to be a salient and effective method of involving community-based education agencies in delivering science experiences that are interesting, educational, and fun, as well as providing an opportunity for leadership and community service to teens.

An Early Survey of 4-H SERIES Services

As 4-H SERIES has evolved, both geographically and in terms of the kinds of programs it influences, the challenge of describing and interpreting the impact of 4-H SERIES has also increased. To date, two kinds of activities have been undertaken to assess the 4-H SERIES program. Extensive surveys were undertaken to estimate the level of participation in 4-H SERIES by 9-to-12-year-olds, teens, and adult volunteers. A second assessment strategy resulted in the development of case studies portraying how the program operates in local settings. A subset of these case studies constitutes the remaining chapters of this monograph. But before introducing the case studies lets look briefly at the survey responses.

In 1992 and 1993, a 13-item two-page survey was completed by 4-H SERIES teen leaders when they completed either: (a) a one-time 4-H SERIES event; or (b) a multiple-session sequence of meetings with a youth group. 4-H SERIES has expanded considerably since these data were collected and the outcomes are unlikely to be descriptive of current use. However, the results give a glimpse of the 4-H SERIES profile in the early stages of program use. After approximately one year of collecting surveys (1992-1993), results were summarized as follows:

◆ *How often were the different curriculum units used?* All existing curricula (six at that time) were used frequently. Snails and Recycle/Reuse were used most frequently (accounting for 28 percent and 19 percent of the total, respectively).

◆ *Were 4-H SERIES events usually single- or multiple-session events?* Approximately 39 percent of the events were single-session and 61 percent occurred over multiple sessions. One third of the events included five or more sessions.

◆ *How much time did 4-H SERIES participants spend in a typical session?* Based on the survey responses, participants typically spent about one hour on 4-H SERIES science activities in one session (49 percent of the sessions were reported to be one hour in length and a substantial portion of sessions (23 percent) were reported to be two or more hours in length).

◆ *What kinds of youth groups participated in 4-H SERIES?* About one half of the groups were 4-H affiliates and about one quarter of the groups were associated with public schools. The remaining groups represented other youth alliances and ad hoc assemblages.

◆ *In what contexts did youths participate in 4-H SERIES events?* About one-third of the 4-H SERIES events were associated with youth camps and about one-fifth with classes. Project or club activities and after school programs each accounted for about one-tenth of events.

◆ *Who led 4-H SERIES events?* 4-H SERIES events were led by teens or teams of teens about 70 percent of the time and by adults about 30 percent of the time.

◆ *How many youths typically participated in a 4-H SERIES event?* Thirty-seven percent of 4-H SERIES events were reported to have between 11 and 20 participants. While few events had less than ten participants, 18 percent had from 21 to 30 participants and 29 percent had more than 30 participants. Or, stated another way, 84 percent of 4-H SERIES events had 11 or more participants.

◆ *What were the ages, genders, and ethnicities of youths participating in 4-H SERIES?* About half of the participants in 4-H SERIES were 9 to 12 year olds. About one-fifth of the participants were under 9 years old and about one-eighth were between 12 and 15 years of age. Slightly more than half (55 percent) of the participants were female. About one-quarter of the participants came from minority groups. Hispanics and Blacks each made up about one-tenth of 4-H SERIES participants.

◆ *How many 4-H SERIES events included community service projects?* Based on 4-H SERIES events with five or more sessions, about one half of the events included community service projects.

◆ *What was the impact of 4-H SERIES on youth?* While it was beyond the scope of the survey to assess the learning outcomes for participants, it was estimated that total 4-H SERIES events up to that time (1992) represented about 470,000 person-hours of hands-on science activities.

In 1993-1994, telephone surveys were carried out in six states and generally paralleled results of the paper and pencil surveys.

Case Studies of 4-H SERIES Implementation

In 1994, a set of case studies was begun to capture, at least in part, how selected aspects of 4-H SERIES were implemented in distinctly different contexts. The diversity of settings in which 4-H SERIES provides services has always been seen as a strength of the program but a bane to assessment efforts with significant resource constraints. Viewing 4-H SERIES as a tool, the case studies were intended to examine several instances of the tool's use. That is, who used the tool, for what purposes, what the context of the use was, how the program was organized to incorporate adults, volunteers, and youths, what activities were undertaken, and what happened as a result of the use? What did 4-H SERIES contribute through its ideas and materials to teens, youth, and adults; to informal and formal science education; to recruitment and preparation of teachers; and to community action and service learning? While these "big" questions are difficult to answer under the best of circumstances, several themes were identified to direct the case studies.

A variety of potential effects of the use of 4-H SERIES have been identified by staff, funders, and participants. From these plausible effects, five major themes were identified and discussed with the 4-H SERIES National Advisory Board to guide the case studies.

The first theme concerned outcomes for participants. When 4-H SERIES is introduced into a context, what are the outcomes for youths, teens, volunteers, program staff, and other stakeholders in the local project? What do participants learn about science content and processes, about themselves, about working with others, and about leadership? Do participants' attitudes toward science, learning, themselves, and career choices change?

The second theme focused on community involvement and community service. 4-H SERIES is designed to have participants apply what they have learned about science to everyday issues and opportunities in their local communities. While not every program implements this aspect of 4-H SERIES, many creative and useful examples of community service and community involvement have been undertaken. Most of the community service projects have involved teens and youth in identifying and addressing a local need, others have created activities and connections with families to learn about science, share youth's accomplishments, or to expand the resources of their communities. What kinds of community action projects have been mounted; who was involved, and what was the result for participants and for the community?

The third theme addressed access to science education and science careers. In many cases, 4-H SERIES has aided agencies in providing services to underrepresented groups in science, especially females, youngsters who are at-risk in formal education settings, and non-English-speaking youth. In what contexts has 4-H SERIES helped to provide access to science education for at-risk youth, what experiences have youth had, and what new opportunities have become available (for example, information about science careers, opportunities to meet and talk with scientists, and visit workplaces)?

The fourth theme concerned characteristics of teaching and learning. This theme examined the extent to which teaching and learning practices have been changed to align more closely with constructivist learning theory. Studies exploring this theme would describe what tasks learners undertake, what tasks "teachers" (*i.e.*, teens and others in teaching-leadership roles) undertake, what "teacher-learner" (learner-learner) interactions are like, what products are produced, and so on. An important element in this theme is the locus of decision making that characterizes the teaching and learning processes. Are learners empowered to be responsible for their own learning, is the balance of power between "teacher-leaders" and learners one-sided or negotiated? Is the learning cycle (the theoretical base of much of 4-H SERIES' instructional design) manifested in the interactions among learners, activities, teacher-leaders, and materials? In each case, this theme should be explored in relation to the pedagogical norms and experiences of the participants.

Finally, the fifth theme looked at impact on the local infrastructure for youth services. In some respects, 4-H SERIES has been a new kind of resource for youth serving agencies, allowing those agencies to work with new populations of youth and volunteers. This aspect of 4-H SERIES is especially germane to, but not restricted to, the 4-H organization. Examining this theme involved looking at the extent and quality of changes in how agency planners selected, provided resources for, implemented and evaluated programs for youth.

While these themes are by no means independent, they served as an umbrella for development and management of a set of case studies. Each case study was intended to illuminate one or more of the themes. Five of these cases, and a chapter on the contributions of community-based science learning experiences, make up the remainder of this monograph.

In Chapter 2, Rachel Davis examines the initiation and implementation of 4-H SERIES in Kentucky. Davis explores the influence of 4-H SERIES on the Kentucky 4-H Youth Development program and asks whether or not 4-H SERIES has supported Kentucky's reform efforts in formal education.

In Chapter 3, Dale Cox describes the impact of participation in 4-H SERIES activities on teens and 9 to 12 year olds in Southern Missouri. This case also examines attitudes toward science and science-related careers among elementary-school age participants.

Teens prepare to test their structures in a simulated earthquake at the 4-H SERIES/SAY training at Skyline High School in Oakland, California.

Chapter 4, by Richard Ponzio, explores impacts on teens of participating in community service projects. From data collected in the 4-HP project in California, teen views about learning and applying science in local action projects are presented.

In the fifth chapter, Charles Fisher and Janice Poda examine the introduction of 4-H SERIES in a state-level precollegiate recruitment and teacher training program. In this case, 4-H SERIES was used as a new component of a much larger, long-standing program in high school teaching academies in South Carolina.

In the sixth chapter, Charles Fisher describes the initiation, development, and impact of Project Excel, a youth and community development project in San Jose, California. In its initial three years, Project Excel used many 4-H SERIES ideas and materials to increase the quantity and quality of hands-on science experiences for elementary school students and to introduce prospective teachers to innovative teaching methods.

The final chapter by Richard Ponzio, describes how the 4-H SERIES project addresses the need for an "educational mosaic" that includes a variety of educational venues and opportunities for children and parents to explore science.

References

Guzzetti, B., Snyder, T., and Glass, G. (1992). Promoting conceptual change in science: Can texts be used effectively? *Journal of Reading, 35,* 542-649.

Karplus, R., Lawson, A., Wollman, W., Appel, M., Bernoff, R., Howe, A., Rusch, J., and Sullivan, F. (1980). *Science teaching and the development of reasoning* (4th Printing). Berkeley, CA: The Regents of the University of California.

Karplus, R., and Thier, H. (1967). *A new way to look at elementary school science.* Chicago, IL: Rand McNally.

Lawson, A., Abraham, M., and Renner, J. (1989). *A theory of instruction: Using the learning cycle to teach science concepts and thinking skills.* Monograph, Number One, National Association for Research in Science Teaching.

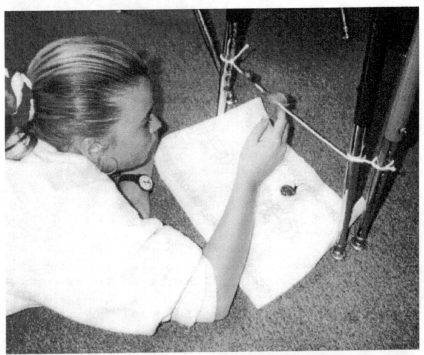

A student at Charles Armstrong Middle School in Belmont, California, watches a snail walk a tightrope as part of a 4-H SERIES "Snail Circus" activity.

Chapter 2

4-H SERIES
Makes an Impact in Kentucky
with Grassroots Adoption

By Rachel Davis

Introduction

This chapter examines the initiation and implementation of 4-H SERIES in Kentucky. The development of this chapter is guided by two questions: What influence has the introduction of 4-H SERIES had on the Kentucky 4-H Youth Development program? And, has 4-H SERIES supported Kentucky's reform efforts in formal education? A short historical overview of 4-H SERIES activities in Kentucky is followed by a brief description of the methodology used to investigate these questions, a summary of results, and a discussion of barriers and openings for future opportunities.

4-H SERIES Activities in Kentucky

The 4-H SERIES program was introduced in Kentucky in 1993. The first state-wide training was initiated by a county 4-H agent. This was unique, because most previous state trainings in Kentucky (for programs other than 4-H SERIES) had been initiated by state 4-H extension specialists working at higher levels in the state 4-H organization. So SERIES was one of the first 4-H programs in Kentucky to be sponsored from the grassroots rather than from the state administration. For this initial training occasion, the 4-H SERIES Regional Leadership Center (in Georgia) provided trainers for six curriculum units. Representatives from 15 counties and nine areas of the state participated, including 14 county extension agents, 14 teens, and 39 adults.

A second state-wide training, conducted primarily by Kentucky 4-H agents who had been trained the previous year, was held in early 1994. The pattern of participation in this second training event was somewhat different. Many counties had realized the value of bringing teams of four to ten people and including teens on each team. At the second training, 21 counties representing eight areas of the state were in attendance. There were nine new agents, seven repeat agents, three program assistants, 42 teens, and 36 adults. The adults were primarily teachers and staff members of various family resource centers. The 4-H SERIES training was designed as a train-the-trainers model and subsequently, participants in the state-wide trainings have conducted other trainings for teens (at 4-H teen retreats or conferences) and school teachers (at in-service events for teachers).

Currently, 4-H SERIES has been operating in Kentucky for two years and, as such, is still a relatively new program. From experiences with dissemination of 4-H SERIES in other states, it takes several years to develop a stable infrastructure for a whole state. Therefore this case study of 4-H SERIES in Kentucky represents an inquiry into the early phase of the implementation process. This case study also examined how this alignment effected the early stages of implementation of 4-H SERIES in Kentucky.

The Kentucky Educational Reform Effort

Kentucky's 4-H program has traditionally had relatively close relationships with the public education system in the state. Even though this kind of relationship is not universal, it is very common in many states. While the success of 4-H programs in Kentucky is not dependent on policies of the state's formal education system, there are occasions when particular 4-H programs are highly compatible with the priorities of the school system. During the period when 4-H SERIES was introduced in Kentucky, the formal education system had been undergoing major change. In 1990, the state legislature passed the Kentucky Education Reform Act (KERA) in an effort to improve the quality and productivity of the system and equalize educational opportunities for all students. KERA's six primary goals for students focus on: (1) use of basic communication and mathematics skills for real-life purposes; (2) understanding of core concepts and principles of science, mathematics, social studies, arts and humanities, practical living skills, and vocational skills; (3) development of self-sufficiency; (4) learning to be a responsible member of a group; (5) reasoning and problem-solving; and (6) drawing connections among various domains of knowledge. Since the 4-H SERIES program has most, if not all, of these components built into its curriculum and experiences, the program is conceptually aligned with KERA's goals.

Methodology

The information presented in this case study was compiled from semi-structured interviews conducted in December 1995 and January 1996. After reviewing the list of extension professionals who had participated in the two initial state-wide trainings, 12 were selected for interviews. The selection process provided data from representatives of most areas of the state and a cross-section of implementation methods that were used. Interviewees were asked to provide a brief summary of their involvement with the 4-H SERIES program and respond to six specific questions about program implementation. Some were asked for names and phone numbers of teachers who had used 4-H SERIES in school settings.

A total of 17 individuals provided responses. One county extension agent responded to the questions by e-mail; five others were interviewed by phone. Three school teachers from three different counties were interviewed by phone. The assistant director of 4-H/Youth Development in Kentucky was also interviewed by phone. Seven teens (from two counties) were interviewed in two face-to-face group interview settings. Interview responses were sorted and examined for possible relationship to the two primary questions of the study. These responses are summarized in the next section along with relevant background information.

Results

Comments from Kentucky Teens Participating in 4-H SERIES

Interviews with 4-H SERIES teen leaders resulted in a positive view of the role of teen participation in 4-H SERIES for youth development, especially development of teaching and leadership skills. Comments from teens about 4-H SERIES:

♦ More options for teaching and doing with kids; different ways of learning.

♦ (4-H SERIES) re-wrote the (local) 4-H program—day camps are huge now, there are more campers in the summer camp program.

♦ ...turned 4-H from contest to education—all I did was enter contests...now I've turned into a teacher...not for the prize but for learning things.

When asked about the responses of others (for example, teachers, principals, other adults and friends), teens were quick to respond:

♦ My Principal was really impressed.

♦ ...we were able to put something back into our home county.

♦ Teachers LOVE it, kids have a ball.

◆ Teachers want to be involved; they get into the activities with the kids.

◆ ...(at Natural Resources Camp) kids were still talking about the day's activities at night.

◆ One teacher said to me (just before a 4-H SERIES lesson) "you'll have trouble with him (referring to a 6th grader); if he doesn't do anything; don't worry about it." Before half the class was over he was talking to me like a brother.

◆ 4-H SERIES is fun to learn and teach.

Participation in 4-H SERIES as a teen leader apparently had a personal effect on teens:

◆ I feel more important, have more self confidence and better communication skills.

◆ I am more comfortable at the front of a room.

◆ It taught me how to work with different people.

◆ I learned how to handle problems in the room.

◆ I'm more knowledgeable about environmental issues on television.

◆ It is a team effort—can't do it all by yourself; give and take.

◆ When kids are enjoying it, it makes you feel good—builds self-esteem.

◆ 4-H SERIES opened my mind, broadened my horizons, opened my eyes to the world.

◆ I like to write poetry; I write about different stuff now, I look past the surface.

◆ I pay closer attention to details and care more about the environment.

◆ I used to go to the answer, now I want to know why.

◆ There is more room, more freedom to mess up and still be OK. We're free to adapt, use our brains.

◆ I have a better concept of teaching others (hands-on versus lecture).

◆ 4-H SERIES helps develop independence; it has brought leadership to us...and taught us how to get to younger kids.

Two teens spoke about differences they noticed in another teen and attributed them to participation in 4-H SERIES:

◆ Mark is involved in what he does; he speaks up. He is more into other people—a leader. He just ran for (and was elected) President of his Senior class; he wouldn't have done that.

Mark's response was that his experiences as 4-H SERIES teen leader gave him the confidence to "step into the unknown and take the risk."

The teens were less clear on how 4-H SERIES aligned with KERA, although in fairness to the teens, their knowledge of KERA may not have been highly detailed. In spite of this, they commented that in 4-H SERIES compared to KERA, you:

♦ Work with groups and hands-on, but you don't have to do all that writing.

They did suggest that in the future 4-H SERIES, like KERA, should be:

♦ More state-wide, more in schools so adults/leaders will use it more.

♦ People just don't know about 4-H SERIES...it should be promoted to teachers.

Comments on 4-H Series from Public School Educators

The Kentucky 4-H program has traditionally had very strong ties to formal education and the public school. Many 4-H programs serve as support for traditional education in the classroom and 4-H SERIES has been no exception. County extension agents reported that 4-H SERIES has primarily been used by either elementary and middle school teachers in their classrooms or as the focus of environmental day camps conducted by 4-H staff and volunteers for elementary school students. However, some school teachers have used 4-H SERIES beyond their regular classrooms in after-school science clubs for 1st through 6th graders and in summer science camps. 4-H SERIES has also been included as a program component in local literacy councils' "read weeks," 4-H fairs, science fair projects, science kits for teachers to check out, and 4-H camp classes.

Mrs. Kleine (Madison Co. 5th grade teacher) originally thought she would use teens from the local high school to conduct 4-H SERIES activities with elementary students in the after-school science club. However, the dismissal times of the local high school and elementary school did not match up. Mrs. Kleine attended the 1994 training and proceeded to use the 4-H SERIES program to supplement her science activities. During the last year she

Student involved in 4-H SERIES "Getting to Know Worms" activity.

has used 4-H SERIES as the basis for her science curriculum.

Teachers who have used 4-H SERIES indicate that its strengths, compared to the more traditional approach, include beginning with students' curiosity (versus just giving them information) and a hands-on approach to science. One high school teacher, serving as a coach/mentor for teens conducting the program in after-school sessions, reported:

◆ The aspects of an after-school enrichment program where older youth experience cooperating on a project that brings school and community together also forwards the efforts of KERA.

Cooperative extension personnel also viewed 4-H SERIES to be highly supportive of state educational policies:

◆ 4-H SERIES is very pro-KERA because of the hands-on, experiential, making-connections type of program it is.

In their interviews, classroom teachers identified additional strengths of the 4-H SERIES program. These strengths included:

◆ All materials needed are listed, easily accessible supplies used for activities, materials are well-written, and goals and expectations for each activity and lesson are laid out so that even someone with limited science background can understand.

◆ Elementary school teachers have limited training and science knowledge. 4-H SERIES is easier for the lay person.

◆ While other classroom science resources (*i.e.,* ACES) have tons of material, there is no book, you can't figure out what is supposed to happen. You need to be more science-oriented to use ACES than 4-H SERIES.

◆ All teachers have been impressed with the format and ease of use. Materials are very teacher-, teen-, and user-friendly. Simple concepts can lead to very complex discussions and exploration, depending upon the user.

These statements provide evidence that 4-H SERIES is filling an unmet need. Several veteran teachers commented:

◆ I have used the 4-H SERIES material more than any in-service training I have received since beginning teaching.

◆ I learned along with the kids.

While teachers were enthusiastic about 4-H SERIES, some interviewees suggested that attendance at local teacher in-services has been hurt because teachers had already earned their required in-service credits, or the local training was not approved for participating teachers to receive professional development credits.

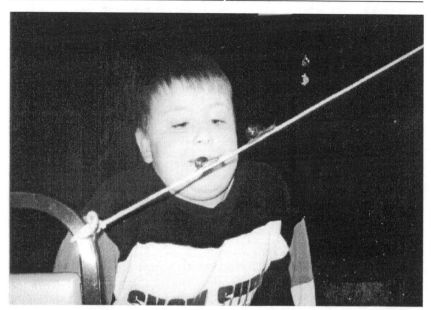

Student testing variables and his hypothesis as part of the 4-H SERIES "Sciencing with Snails/Snail Circus" activity.

Comments on 4-H Series from County Extension Agents

County agents used 4-H SERIES in a variety of ways and for a variety of purposes. In addition to facilitating implementation of 4-H SERIES in formal education settings, agents used 4-H SERIES as a resource for 4-H school club meetings and, in some cases, in after-school child care programs. According to the county extension agents interviewed, the introduction of 4-H SERIES in Kentucky produced many benefits for 4-H programming efforts. They indicated that the program:

◆ Reinforces [the] hands-on learning motto.

◆ [Makes it] easy to keep students interested in the material—they love it!

◆ [Is an] opportunity for teens to teach subject matter they could handle and enjoy.

◆ [Is] user friendly for teens; they can use other curricula [from the 4-H SERIES program] even if they have not been trained in it.

◆ Has improved the quality of 4-H projects.

On a larger scale it has been an excellent marketing tool by strengthening county 4-H agents' reputations as resource persons for teachers. One agent who has trained 60-70 teachers in her local county described 4-H SERIES as:

◆ The most successful ever in terms of leader training.

◆ The only 4-H curriculum packaged to teach science.

◆ An excellent example of quality 4-H informal educational methods.

County extension agents were also positive about use of 4-H SERIES in both school and after-school settings:

◆ Teachers have been bootlegging the curriculum and sharing it with their colleagues.

◆ One group of 4-H teens taught snailing to a group of at-risk youth at our family resource center. When the class started, the students had very tough, "who wants to fool with snails," "oh, how dumb," "yuck, touch a snail" attitudes. Before the class was over, they were naming their snails and wanted to take them home with them!

In Garrard County, 4-H SERIES was a vehicle to involve previously unreached teens in 4-H. Their high school science teacher served as a volunteer coach with the local extension agent. As a result of leading 4-H SERIES units with youth after-school:

◆ These teens had a new perspective on teaching—they experienced the frustration of not being prepared and having kids that weren't focused on the activities. They also had to count on each other as team members and were angry when some teens didn't show up for the planning meetings or activity sessions with the younger kids.

◆ They found that 4-H SERIES seemed to provide a powerful opportunity for responsibility—the teens involved in this activity were more dependable and showed more initiative than teens involved in other 4-H activities at the same time. This was evident with teens adjusting work and other extra-curricula activities in order to fulfill their time commitments to 4-H SERIES.

Grassroots Adoption of 4-H Series Within 4-H in Kentucky

In the two years that 4-H SERIES has been available in Kentucky it has spread by local adoption. That is, an individual has taken responsibility for setting up training and developing a local 4-H SERIES science program. This process has generated a lot of 4-H SERIES activities across the state but the program has not been adopted at the state level. Professionals working within Cooperative Extension were asked about this situation. Some responses suggested that 4-H SERIES appealed to the particular personalities of individual agents but that the program had not yet attracted sufficient support to warrant state-wide adoption. Others suggested that a large number (and proportion) of newly hired agents were somewhat overwhelmed by the demands of the job and were not in a position to advocate specific programs. When asked directly about a broader mandate, county agents commented:

♦ I already have more to do than I can manage.

♦ Other agents don't feel like it is very important.

♦ Those (agents) that do know about it like it very much.

♦ Agents may know that it deals with science, they hear good things, think they should pay more attention to it, but haven't.

♦ Since there was no state mandate of this program, the understanding and knowledge of the program has been hit or miss.

An interview with Kentucky's assistant director of 4-H Youth Development provided another context in which to view the adoption process of 4-H SERIES. Traditionally Kentucky 4-H has focused on particular programs and events with direction given by the state 4-H office through an adoption process. When a program has achieved state-wide adoption, there has been an expectation that events related to that program will occur in all counties. According to the assistant director, the state is:

♦ Transitioning to focus on longer-term involvement. The state will identify priority areas. In the last few years programming decisions have been shifted to the county level. The state 4-H Youth Development Department will focus on providing curriculum, resources and support in three priority content areas: (1) workforce preparation; (2) science and technology; and (3) environmental stewardship. Professional staff working with county 4-H councils will need to determine local needs and priorities and select from a set of state-wide program resources.

While 4-H SERIES implementation has not been mandated, state 4-H officials have encouraged and supported continued training efforts. State 4-H leaders reiterated several potential benefits of participation in 4-H SERIES to the Kentucky 4-H Program. Interviewees described 4-H SERIES:

♦ A resource for classroom clubs—more hands-on activities; you don't have to know much about experimentation, you just have to do it.

♦ It demonstrates to teachers hands-on and discovery learning.

♦ It also strengthens Kentucky 4-H's position as a partner in education with the newly-formed Rural Systemic Initiative to improve the teaching of science and math in 33 Appalachian counties.

♦ 4-H SERIES is a vehicle to get other university faculty more interested in 4-H and take responsibility for outreach efforts.

In summary, it was clear that all individuals interviewed liked 4-H SERIES. Teens liked the "variation" and "activities." They also liked "teens and adults working together—team building." Other interviewees noted that 4-H SERIES "helped the status of 4-H" and "covers lots of material." Adult users liked the "neat package," the "curriculum," the fact that is was "easy to use" and was a "vehicle for teens." All groups "would like more units (for example, geology)."

Discussion

These comments are taken to be evidence of benefits to teens from participation in 4-H SERIES and therefore an indication that 4-H SERIES has had a positive influence on youth development in the state. There seems to be little doubt that 4-H SERIES fills a need for teachers in their efforts to provide experiences for students that fulfill KERA's goals.

While 4-H SERIES has not yet obtained state-wide adoption, state officials assert that the program provides a solid, sustainable base for future programming. Even if there were no more 4-H SERIES training events, 4-H SERIES would continue to make an ongoing impact in individual counties for years to come.

Two strengths of the 4-H SERIES model are cross-age teaching and application of learning to community service. While cross-age teaching has been implemented, widespread use of the community service component of 4-H SERIES has yet to receive adequate attention. Teens have provided leadership in day camp and 4-H camp environments; however for the most part these involved a sampling of the activities from different units presented to small groups of different elementary students. This has left little opportunity for younger youth to develop on-going relationships with the teens. In Garrard County, teens worked with youth at the after-school child care programs. They were able to know some of the youth individually by name. In this setting there was still some inconsistency because not all children attended the after-school program every day.

Most of the curriculum units have been taught by teachers in the classroom during the school day. The logistical details of getting teens out of their high school classes and transported to elementary schools to lead activities have so far proven to be a barrier to cross-age learning in the classroom. However, given the positive reports of teens in this small sample, there appears to be excellent potential for extended use of 4-H SERIES in Kentucky, especially for youth development and leadership training.

The implementation of 4-H SERIES in classroom environments has, for the most part, not included the community service component of applying the knowledge learned in the unit. Teachers need assistance in designing and implementing real-life opportunities for youth to apply the knowledge learned. Perhaps school teachers and 4-H extension staff could design a partnership for delivery that would have teachers providing primary leadership in teaching the curriculum lessons and extension staff providing leadership and organization in the community service component. The introduction of 4-H SERIES in Kentucky is still in the early adoption stages. There are several aspects of 4-H SERIES of which we haven't taken full advantage. We have been thinking about the 4-H SERIES program in terms of our old paradigm: that is, 4-H SERIES is one of many programs that we can choose to implement. Another possibility would be to explore how the 4-H SERIES program could be a primary vehicle/context for accomplishing many of our goals for youth development. Leadership development training for teens is built into the 4-H SERIES model, as are life-skills development for both older and younger youth, science literacy, employability skills, and community service opportunities. Using 4-H SERIES as a context for 4-H programming would provide connections among program components that currently are missing.

While Kentucky 4-H has traditionally been very involved in support of formal education efforts at the local level, it has not received recognition at the state level. 4-H SERIES could be a vehicle to establish a stronger partnership between formal and nonformal education programs like 4-H. Kentucky high schools will soon be expected to require community service of graduating seniors. Formal education officials are at a loss as to how to organize and manage these experiences for their students. This is a need that 4-H staff can fill, and 4-H SERIES can provide a pre-packaged, proven vehicle.

In summary, the 4-H SERIES program has made substantial progress in the state. Continued growth in the use of 4-H SERIES seems inevitable. This 4-H science education program has yet to reach its full potential in Kentucky.

Student participating in 4-H SERIES "Oak Woodlands/Wood-peckers and the Oaks" activity learns the importance of adequate food supply in order for woodpeckers to survive and reproduce, as well as that more than one wildlife species often uses the same food source in the same habitat.

Chapter 3

SERIES Implemented in a Rural 4-H Setting:

The Case of Lamar, Missouri

By Dale Cox

Introduction

While 4-H SERIES is a nationally disseminated science program, its implementation varies considerably with local conditions in various parts of the United States. For example, there are distinct differences in the relationships between 4-H programs and public schools from state to state. In many areas of the country, 4-H programs work very closely with schools while in others the two organizations are relatively separate. This chapter explores a case in which 4-H and local schools work together to use 4-H SERIES as a leadership development tool for teens and at the same time provide hands-on science opportunities for elementary school students and valuable training in contemporary science activities for their teachers.

This case study tells part of the story of 4-H SERIES in the southwest corner of Missouri. The data describe responses to 4-H SERIES among teens, teachers, and elementary school students at Lamar High School and Lamar Elementary School. The work on which this case is based began with a focus on the attitudes of elementary school students toward science. The question driving the inquiry was: "Does experience in 4-H SERIES science activities over the course of a school year influence attitudes toward science among students in grades 2 through 4?" From this beginning, the study soon branched out to include other information about the local context.

With support from the local 4-H office, a cadre of teens from Lamar High School was introduced to 4-H SERIES at the North Central States training session in March 1992. After their initial training, teens presented 4-H SERIES programs to 4-H members during the annual 4-H camp. A second general training was held in March of 1993. In the fall of 1993 a third training was conducted in Mt. Vernon, Missouri, especially for public schools by high school students from Lawrence, Barry, and Greene Counties in southwest Missouri. At that time, additional students from Lamar High were trained as well as students from a number of other public schools and Christian schools, and several home schoolers.

From the start, the group from Lamar, compared to other groups, has been the most active. The team received leadership from Mrs. Laurann Robertson, a high school chemistry teacher, and Juston Puckett, a high school student coordinator. The local elementary school principal, Jan VanGilder, was also very supportive and cooperative.

Teenage 4-H SERIES trainers worked with the elementary teachers to develop a schedule for presenting 4-H SERIES lessons to the elementary students. The elementary teachers scheduled the 4-H SERIES activities as part of their regular curriculum either as an introductory unit or as a summary. At the time, 4-H SERIES had six curricular units and teachers chose to have from one to six of them taught in their classrooms. Fourteen teachers participated in the initiative and 12 had six or more presentations in their classes. The remaining two teachers had one presentation each in their classes. Data presented later in this case study come from the 12 classes where 4-H SERIES was used relatively frequently.

Elementary School Students' Attitudes Toward Science

The central question in this case study concerned the effect of participation in 4-H SERIES on attitudes toward science among the elementary school students. To assess these attitudes, elementary school students were given the Draw-a-Scientist Task (DAST) and a semantic differential task in November 1994 before they had engaged in 4-H SERIES activities and again in May 1995 after they had had considerable experience in the 4-H SERIES program.

The Draw-a-Scientist Task

In the DAST, students are asked to draw a representation of a scientist on a single sheet of paper. The drawings that students produce are analyzed for a number of characteristics that have been related to attitudes to science (see for example, Chambers, 1983; Mason, Kahle, and Gardner, 1991). The fall administration of the DAST was carried out by the high school students who would later be presenting 4-H SERIES. Students in grades 2, 3, and 4 produced drawings. The 4-H SERIES program was being used in all grades in the school but since grades 5 and 6 had already started, no DAST data were collected in these grades. All of the pretests were administered on the same day. Students recorded their name, grade, gender, and classroom teacher on sheets of paper provided by the administrators and then drew their scientists. The high school students had been previously instructed not to give too many instructions or hints about what was expected. In the spring, the same high school students collected post tests from the elementary students. With the exception of one class that had a scheduling conflict, all DAST spring data were collected on the same day.

Three types of coding were completed. First, each drawing was coded for the gender of the scientist represented. Responses were coded as male, female or other. The "other" category included cases where no human or other live entity was depicted, where the gender of the scientist could not be determined, and where both a male and a female depiction was included in the same drawing.

The second and third types of coding were adapted from Mason, Kahle and Gardner (1991). The second type of coding involved identifying typical attributes of science and scientists in the drawings. Following Mason *et al.*, 11 attributes or indicators were coded: (1) laboratory coat; (2) eyeglasses; (3) facial hair; (4) symbols of research (*i.e.*, test tubes, flasks, Bunsen burners, microscopes, etc.);

Student investigates the physical and behavioral attributes of a snail as part of the 4-H SERIES "Sciencing with Snails" activity.

(5) symbols of knowledge (*i.e.,* books, filing cabinets, etc.); (6) signs of technology and products of science (*i.e.,* solutions in glassware, machines, etc.); (7) captions; (8) male figure; (9) signs and labels; (10) pencils and pens; and (11) unkempt appearance. Each response was coded for these indicators and then the total number of indicators for each drawing was calculated.

The third type of coding was done globally for each drawing. Following Mason *et al.*, drawings were coded as sinister (involving threat, violence, or apparent destruction), eccentric (unusual, weird, or goofy characterization of the scientist), neutral, or other (uncategorizable in the other three categories). At a later stage of the analysis, the sinister and eccentric categories, neither of which occurred frequently in these data, were combined.

The coding was done by three persons: myself, a college student, and an

Table 1

Gender of Scientists Drawn by SERIES Participants at Lamar Elementary School (from the draw-a-scientist task)

Student Gender	Grade Level	N	November 1994			May 1995		
			Male Fig.	Female Fig.	Other Fig.	Male Fig.	Female Fig.	Other Fig.
Males	2	39	25 (64)	0 (00)	14 (36)	24 (62)	0 (00)	15 (38)
	3	44	28 (64)	2 (4)	14 (32)	30 (68)	5 (11)	9 (20)
	4	39	28 (72)	3 (8)	8 (20)	28 (72)	0 (00)	11 (28)
Females	2	25	5 (20)	13 (52)	7 (28)	6 (24)	13 (52)	6 (24)
	3	41	16 (39)	18 (44)	7 (17)	13 (32)	24 (59)	4 (10)
	4	30	9 (30)	17 (57)	4 (13)	3 (10)	23 (77)	4 (13)
Total Males		122	81 (66)	5 (4)	36 (30)	82 (67)	5 (4)	35 (29)
Total Females		96	30 (31)	48 (50)	18 (19)	22 (23)	60 (63)	14 (15)
Total		218	121 (56)	53 (24)	54 (25)	104 (48)	65 (30)	49 (22)

Entries in the body of the table represent frequencies; numbers in parentheses are percents.

elementary school teacher. As differences in coding came up they were discussed and resolved. Because of the teacher's experience with the students in school, her advice was often taken. Only those students who had completed both pretest and post-test drawings were included in the data coding and analysis. Four hundred and thirty-six drawings were coded; two for each of 218 students (122 males and 96 females).

Results

Results are presented for grades 2, 3, and 4 for boys and girls on each of the three coding rubrics.

Gender of scientists drawn by the students. Table 1 shows the gender of scientists drawn by the students. From the bottom row of Table 1, note that from November to May students drew 8 percent fewer male scientists (56 percent male in November and 48 percent male in May). Over the same period, drawings of female scientists increased by 6 percent and drawings that could not be categorized as either male or female decreased slightly (3 percent). These percentages are clarified somewhat by examining the body of Table 1 where results are broken down by gender and grade level of students. The percentages for boys are practically identical over the six-month testing period. Both before and after participation in 4-H SERIES, boys tended to draw scientists as male figures about two-thirds of the time. Boys drew scientists that could not be coded as either male or female about a third of the time and occasionally (4 percent of the time) they drew scientists as female figures. These results were very similar for boys regardless of their grade level (with boys in grade 2 drawing figures that were neither male nor female slightly more often).

While the data are not entirely consistent, there appear to be solid trends in the drawings made by girls. Averaged over the three grade levels, girls drew fewer male scientists and more female scientists after participating in 4-H SERIES than before (male figures went from 31 percent in fall to 23 percent the next spring while female figures went from 50 percent to 63 percent). When grade level is considered, it is clear that no change occurred in the drawings for girls in grade 2, but in grades 3 and 4, there were marked and consistent shifts. From these data it appears that girls attitudes toward science, as represented in the depiction of scientists as female, are more positive after participation in 4-H SERIES.

Number of science stereotypes appearing in the drawings. When the students' drawings were coded for the number of stereotypic elements they contained, one trend appeared. For both boys and girls, the number of

Table 2

Average Number of Stereotypic Indicators in Elementary School Students' Drawings (from the Draw-a-Scientist Task)

Student Gender	Grade Level	# of Students	November 1994	May 1995
Males	2	39	2.8	2.7
	3	44	3.0	3.1
	4	39	3.8	3.7
Females	2	25	2.3	2.7
	3	41	2.9	3.5
	4	30	4.0	3.7
Total Males		122	3.2	3.2
Total Females		96	3.1	3.3
Total		218	3.1	3.2

indicators increased with grade level. It appears that students are learning about these stereotypes and include more of them in their drawings as they grow older (and presumably accumulate experience with science and scientists). Given the age of the students, this may be a positive sign. In adults, the presence of stereotypes is interpreted as leading to a negative attitude toward science—but with young children this may not be the case. Table 2 presents results for students in this study.

Global coding of students' drawings. The global coding scheme centers on the number of sinister or eccentric figures drawn by students. The notion behind the rubric is that a reduction in sinister and eccentric drawings represents a positive shift in attitude toward science. In the data gathered at Lamar Elementary School, there were no obvious trends in the percentages of sinister and eccentric elements in students' drawings, either for boys or girls, regardless of grade level. Data for the global coding are included in Table 3.

Table 3

Sinister, Eccentric, and Neutral Scientists Drawn by SERIES Participants at Lamar Elementary School (from the Draw-a-Scientist Task)

Student Gender	Grade Level	N	November 1994			May 1995		
			Sinister/ Eccentric	Neutral	Other	Sinister/ Eccentric	Neutral	Other
Males	2	39	0 (00)	32 (82)	7 (18)	1 (3)	28 (72)	10 (26)
	3	44	8 (18)	34 (77)	2 (5)	4 (9)	28 (64)	12 (27)
	4	39	3 (8)	33 (85)	3 (8)	3 (8)	25 (64)	11 (28)
Females	2	25	1 (4)	23 (92)	1 (4)	0 (00)	20 (80)	5 (20)
	3	41	3 (7)	36 (88)	3 (7)	8 (20)	32 (78)	1 (2)
	4	30	2 (7)	26 (87)	2 (7)	0 (00)	28 (93)	2 (7)
Total Males		122	11 (9)	99 (81)	12 (10)	8 (7)	81 (66)	33 (27)
Total Females		96	6 (6)	85 (89)	6 (6)	8 (8)	80 (83)	8 (8)
Total		218	17 (8)	184 (84)	18 (8)	16 (7)	161 (74)	41 (19)

Entries in the body of the table represent frequencies; numbers in parentheses are percents.

There appear to be trends in the "other" category, especially for boys. However, no clear interpretation of these trends is suggested here. *Summary of results on Draw-a-Scientist Task.* Of the three coding rubrics, the largest differences in students' drawings from November to May occurred in the gender attributed to scientists. While boys consistently drew scientists as male, indicating a positive attitude to science, girls increased the frequency with which they depicted scientists as female, indicating an increasingly positive attitude toward science. The numbers of stereotypic indicators increased for both boys and girls as a function of grade level but apparently not with participation in 4-H SERIES. The global coding showed no consistent increase or decrease in depiction of scientists as either sinister or eccentric.

In addition to the coding schemes there were some indications in the content of the drawings that students' notions of scientists and science were changing as they participated in 4-H SERIES activities. The prime example in this regard concerns the setting that is most common for scientists. Compared to the pretest, post-test drawings more often depicted scientists out of the laboratory and into the larger world. Students showed their scientist doing "outside" work. Digging dinosaur bones was a popular activity. A few students tried to depict one or more of the scientific processes—like observation, communication, and organizing—in their drawings. These processes are explicitly included in the 4-H SERIES curriculum.

The Semantic Differential Task

The semantic differential technique is a long-standing procedure that has been used, among other things, for assessing attitudes (Osgood, Suci, and Tannenbaum, 1957). In the current study, grade 3 and 4 students were given 17 adjective pairs (separated by 7-point scales) with which to describe the concept "science." The adjectives, balanced for polarity, were presented on one page and the task was administered in school classrooms in the fall and spring of the school year (before and after participation in 4-H SERIES). Four fourth-grade and three third-grade classes participated. Results for the various items are displayed in Tables 4 through 7. While the items have been grouped in the traditional evaluation, potency, and activity categories, they have not been combined to yield scale scores.

Perusal of the evaluation items (see Tables 4 and 6) shows considerable inconsistency across grade levels and gender. There was moderate consistency for boys and girls in both grades on two items. Students described science as more "valuable" and "harder" after participation in 4-H SERIES. For the potency and activity items (see Tables 5 and 7), there were no items that showed consistent direction for both gender and grade.

Table 4

Attitude toward Science among Fourth-Grade Students at Lamar Elementary School (from semantic differential)

Gender	Date	N	Good-Bad	Valuable-Worthless	Pleasant-Unpleasant	Kind-Cruel	Fair-Unfair	Honest-Dishonest	Easy-Hard	Interesting-Boring
							Evaluation Items			
Male	Nov.94	44	2.2	1.7	2.2	1.8	2.1	1.7	-0.1	2.3
	May 95	44	1.5	2.0	1.2	1.2	1.4	1.7	-0.5	1.9
Female	Nov. 94	28	2.2	2.4	1.5	2.4	2.4	1.9	0.2	2.4
	May 95	28	2.1	2.3	1.7	1.0	1.5	2.1	-0.2	2.0
Total	Nov. 94	72	2.2	2.0	2.1	1.7	2.3	1.8	0.0	2.4
	May 95	72	1.8	2.1	1.4	1.1	1.4	1.8	-0.4	2.0

Table 5

Attitude toward Science among Fourth-Grade Students at Lamar Elementary School (from semantic differential)

Gender	Date	N	Fast-Slow	Safe-Dangerous	Rugged-Fragile	Strong-Weak	Hot-Cold	Noisy-Quiet	Active-Calm	Outside-Inside
				Potency Items				**Activity Items**		
Male	Nov.94	44	0.5	-0.2	-1.0	1.4	0.8	-0.2	1.3	-0.8
	May 95	44	-0.5	-0.6	0.2	1.0	1.2	0.5	1.4	-0.6
Female	Nov. 94	28	-0.5	0.1	0.5	1.5	0.9	0.8	2.1	-0.6
	May 95	28	0.6	-0.1	-0.4	1.4	1.0	0.2	2.1	-0.4
Total	Nov. 94	72	0.1	-0.1	-0.4	1.4	0.9	0.2	1.6	-0.7
	May 95	72	0.0	-0.4	0.0	1.2	1.1	0.4	1.7	-0.5

Data were collected in a total of four classrooms using a 7-point differential (-3 to +3)

Table 6

Attitude toward Science among Third-Grade Students at Lamar Elementary School (from semantic differential)

Gender	Date	N	Evaluation Items							
			Good-Bad	Valuable-Worthless	Pleasant-Unpleasant	Kind-Cruel	Fair-Unfair	Honest-Dishonest	Easy-Hard	Interesting-Boring
Male	Nov. 94	33	1.3	1.5	1.0	0.8	1.2	1.3	-0.8	1.8
	May 95	33	1.9	2.1	1.0	1.3	1.3	1.3	-1.4	2.2
Female	Nov. 94	31	1.9	1.8	1.8	1.6	1.7	1.5	0.5	2.3
	May 95	31	2.1	2.2	1.8	1.6	1.5	1.7	-0.1	2.5
Total	Nov. 94	64	1.6	1.7	1.4	1.2	1.4	1.4	-0.2	2.1
	May 95	64	2.0	2.1	1.4	1.5	1.4	1.5	-0.8	2.4

Table 7

Attitude toward Science among Third-Grade Students at Lamar Elementary School (from semantic differential)

Gender	Date	N	Potency Items					Activity Items		
			Fast-Slow	Safe-Dangerous	Rugged-Fragile	Strong-Weak	Hot-Cold	Noisy-Quiet	Active-Calm	Outside-Inside
Male	Nov. 94	33	0.1	-0.4	-0.2	2.0	0.6	-0.1	0.6	-0.8
	May 95	33	-0.2	-0.2	-0.2	1.6	0.8	1.0	2.0	-0.4
Female	Nov. 94	31	0.5	0.9	-0.6	1.5	1.0	-0.7	1.1	-0.8
	May 95	31	0.1	0.0	0.2	1.4	0.5	0.4	1.4	-0.5
Total	Nov. 94	64	0.3	0.3	-0.4	1.7	0.8	-0.4	0.8	-0.8
	May 95	64	-0.1	-0.1	0.0	1.5	0.7	0.7	1.7	-0.5

Data were collected in a total of four classrooms using a 7-point differential (-3 to +3)

Parents, Teachers, and Teens Assess 4-H SERIES Program

In addition to examining students attitudes toward science, the Lamar case study explored the responses of several other stakeholders in the project. The next section of the study reports on each of these groups in turn.

Views of the Parents of Elementary School Students

Parents were asked to respond to a short survey about the Lamar project. The survey was taken home by students near the end of the school year. Based on 116 surveys returned by parents (23 from parents of grade 2 students; 54 and 39 from grades 3 and 4, respectively), parents made the following comments about the program. A high proportion of parents reported that their children talked with them about science experiences in school (78 percent in grade 2, 93 percent in grade 3, and 79 percent in grade 4). About half of the students talked directly about 4-H SERIES science activities with their parents (43 percent in grade 2, 40 percent in grade 3, and 59 percent in grade 4). When parents were asked if their child's interest in science had increased over the school year, about half responded that it had (47 percent in grade 2, 34 percent in grade 3, and 57 percent in grade 4). It is interesting to note that the percentage of parents reporting increased interest in science by their child is almost the same as the number of parents reporting that their child talked specifically about 4-H SERIES at home. These responses from parents were interpreted as a positive assessment of 4-H SERIES science activities.

Views of Elementary School Teachers

Classroom teachers provided feedback on the program in two ways: through comment cards given to teen leaders after individual lessons in the elementary school classrooms; and during interviews about the project conducted near the end of the school year.

Classroom teachers were asked to respond to questions on a 3x5 card each time a 4-H SERIES presentation was given in their classrooms by teens. These questions were developed by the teens and dealt primarily with the teaching skills of the teens while in the elementary teacher's classroom. Classroom teachers responded to the questions and returned the cards to the teens at the completion of the session. Typical comments included:

♦ Thank you for explaining 4-H SERIES to us. That's the first time any group has done that. You seemed well prepared and related well to the children. They loved the activities.

♦ This was a great program—I know the kids liked it. It goes well with the 3rd grade material.

Most other teacher comments were also very favorable to the program.

When teachers were interviewed by 4-H staff near the end of the school year, there was additional positive feedback for the program. Predictably, there were some scheduling problems reported but all teachers viewed the program to be a useful contribution to their students. Several comments indicate that teens first had to overcome some expectations as they entered the elementary school classes:

Students explore engineering and architecture by building structures using toothpicks, straws, and marshmallows on a foundation of jello.

◆ The first year we thought "Oh boy, kids [teens] are coming to our class and they are just going to diddle around for an hour and waste our time and they probably won't know their material," but they presented it right.

◆ The first year we just wanted to sign up and get it over so that we could go on, but the next year we were ready to sign up because they had done it so well.

Teachers reported that their students retained and used some of the vocabulary about scientific processes:

◆ Well, I'd say some of my brighter students picked up on the terminology. We heard it the rest of the week. They would say "Yeah, remember they wrote it on the board. It was observing."

Although academic achievement was not measured as part of this case study, teachers reported that 4-H SERIES activities helped students make connections with their other science activities.

When teachers were asked if they would do the 4-H SERIES activities themselves, they said they probably would not because of lack of preparation time, storage space, and resources to purchase the materials. In other comments, a second grade teacher mentioned that her students especially liked the high school students coming in:

◆ They could be talking about anything. It was a social thing.

Another teacher reported:

◆ I have a learning disabled student that I haven't gotten much out of all year, but he enjoyed making the "bug" [part of a 4-H SERIES activity on pests] and telling what it was and the enemies of the bug. The hands-on things are great. The kids loved it.

Views of the High School Teens

Near the end of the school year, teens were interviewed about their participation in 4-H SERIES. When asked how participation in 4-H SERIES had affected their career choices, most claimed to have made choices before being engaged in 4-H SERIES. Others responded with:

◆ It has made me think more seriously about teaching.

◆ I would absolutely be a teacher in a second. I have also developed many management skills.

To the question, "Has 4-H SERIES helped you investigate other areas of interest in science?" teens gave almost unanimous agreement and mentioned all of the 4-H SERIES curriculum units. When asked what was most exciting about the presentations, teens reported "teaching kids," "working with kids" and "the hands-on stuff." The least enjoyable part of the experience they reported to be, "Kids who won't obey and listen to directions" and, in rare cases, a teacher who wouldn't cooperate. When asked for suggestion for improvement of the program, they mentioned "more [curriculum] activities," "more work with classroom teachers," "go to the class before [presenting] and hang up posters," and spending more time in the elementary school classrooms. When teens were asked how their high school teachers responded to their missing high school classes, they reported, "they usually understand as long as we keep up [with high school work]," and, "they liked it and felt it was a good experience for us."

To provide another view of teens participation in 4-H SERIES activities, Mrs. Robertson, the high school coordinator for the program, was asked to contact other teachers in the high school and report on teens' grades and attitudes in high school classes. She compiled responses for 20 high school students and only one student's grade had decreased. Even in that case, there were extenuating circumstances since the student had missed several days of school because of illness. High school teachers reported that the attitudes of 4-H SERIES participants in their classes were good to excellent. The report did not indicate how many class sessions teens had missed but, since most teens presented 15 to 20 lessons and each presentation required some travel and preparation time, the numbers of missed classes could have been substantial. In spite of this, 4-H SERIES teens were in good standing and succeeding in their high school course work.

Conclusion

The 4-H SERIES program generally worked well, both for teens in Lamar High School and for younger youth in Lamar Elementary School. The program was practical, in that it could be implemented within the usual conditions of formal classroom education. Teachers, parents, and teens were positive about its effects on science learning and youth development. There was some evidence that among elementary students, participation in 4-H SERIES was correlated with increases in favorable attitudes toward science for females.

References

Chambers, D.W. (1983). Stereotypic images of the scientist: The draw-a-scientist test. *Science Education.* 67, 255-265.

Mason, C., Kahle J., and Gardner, A. (1991). Draw-a-scientist test: Future implications. *School Science and Mathematics*, 91(5), 193-198.

Osgood, C., Suci, G., and Tannenbaum, P. (1957). *The measurement of meaning.* Urbana, IL: University of Illinois Press.

Students use recycled pegboard to create a display on recycling.

Students admire the community gardens they created, as part of their study of agriculture, to help feed the homeless in their neighborhood.

Chapter 4

Science-Based Community Service Projects:
A Potent Context for Learning

By Richard Ponzio

Introduction

Since its inception, 4-H SERIES has included community service projects in its design for developing the cognitive and leadership skills of teens. By combining a variety of important learning conditions, community service projects appear to be particularly potent contexts for learning. The projects involve teens in action-oriented groups working on self-selected community problems with the support of a mentor or coach. The projects are invariably complex enough to require creative problem solving along the way. This chapter provides an example of how such projects can be organized under a larger umbrella structure and explores the impact on teens of participating in projects.

With support from the Hewlett-Packard Company, the National 4-H SERIES team launched an initiative referred to as the 4-HP project in September 1993. In the course of the 4-HP project, 26 science-based community service projects were undertaken by 170 California teenage volunteers who, in turn, recruited more than 5,000 younger youths as participants. Each of the 26 teen-led community projects used activities and materials developed by the 4-H SERIES project as their underlying resource for science inquiry.

This chapter examines the impact of participation in science-related, community service projects on teens and, in some cases, on the communities in which they completed their work. Did the projects benefit the communities, teens, and younger children who participated? The intended outcomes and experiences for teens included:

◆ Using an inquiry model to identify and frame authentic community problems related to the science themes in the 4-H SERIES curricula;

◆ Preparing a goal statement, proposing a plan of work to solve the identified problem, and planning an evaluation of the desired results;

◆ Recruiting and preparing younger youth to understand the science underlying the problem at hand and enrolling them to participate in the project;

◆ Working effectively with an adult coach; and

◆ Developing a portfolio to document and summarize their overall project design, problem solving, implementation efforts, and accomplishments.

In our examination of the 4-HP project, we sought evidence that teens experienced these processes. We also tried to assess the effectiveness of teens as instructional leaders for the 9-to-12-year-old participants in the inquiry-based science activities. In the last decade, there has been renewed interest in community service projects as a key element in the educational development of productive workers and citizens (see for example Reich, 1991; 1994). Educators, youth advocates, economists, and politicians alike have championed community service projects as a mechanism for leadership development, instilling a sense of community responsibility through volunteerism, and preparing for 21st century careers.

The 4-HP Project

The 4-HP project got under way with two 4-H SERIES training conferences that were attended by 170 teens and 55 adult coaches from a wide variety of localities in California. During the training sessions both teens and adults received extensive training in the purposes and use of 4-H SERIES curricula. Subsequently teens were invited to propose a science-based community service project that could be implemented in their home communities. The request for proposals offered reimbursement for travel, materials, and supplies needed to carry out the proposed project. No money was offered for salaries. Funding for the teen proposals was supplied by a generous grant from the Hewlett-Packard Company. A 4-HP project advisory board was set up to receive and review the proposals. Approximately 30 proposals were submitted by teens for review. The advisory board reviewed and responded to all proposals, in every case asking for clarification, justification, or revision. Twenty-six projects were eventually funded; awards ranged from $250 to $2,700. The 26 projects involved 169 teens and 26 adult coaches and eventually engaged more than 5,000 nine to 12 year olds and approximately 58,000 community residents in one or more of the projects' activities (4-H SERIES project, 1995).

Each of the community service projects was designed by the teen leaders themselves. In most instances two to five teens worked as a team, exploring their community's needs with regard to a specific 4-H SERIES curriculum subject matter. They then met with their adult coach, in many cases one of their high-school teachers, and began presenting 4-H SERIES activities to a group of 9-to-12-year-old children. As part of the 4-H SERIES curriculum, they began planning a practical project that would involve working with their 9-to-12-year-old participants in solving an identified community need.

Although the 4-HP projects came from a variety of urban, rural, and suburban settings as well as from a variety of groups (for example, 4-H clubs, community agencies, continuation high schools), the proposals had several features in common. Each had identified a recognized community need that included developing higher science literacy levels among local nine to 12 year olds and proposed a set of actions to mitigate a particular science-related problem or circumstance (examples included watershed restoration, disaster preparedness, chemical awareness and safety, oak woodland habitat restoration, beach cleanups, and so on). No two projects approached any one issue in quite the same way. Each proposal included an analysis and plan to address the identified community problem. This included a description of the problem or issue and, in most cases, corroboration by an expert or local agency official, an innovative approach to solving the problem by taking some form of appropriate action involving the younger (9-to-12-year-old) participants, a well-defined scope of work, an explanation and justification for all budget items, and a plan for how they would assess the impact on their actions on the identified problem. Each project also constructed and submitted a portfolio documenting the progress and outcomes of the project.

Brief Descriptions of Science-based Community Service Projects

Teens undertook a remarkable variety of projects. The following examples provide a sense of the range of topics and strategies that teens proposed and later implemented in their projects:

◆ Teens in Santa Barbara worked with the U.S. Forest Service to gain a greater understanding of the value and importance of native foliage as wildlife habitat in rural landscapes. The teens and youth participants initiated a local action project by collecting a variety of native plant seeds, planting and caring for the seedlings in a green house setting, then planting the young plants in fire- or flood-damaged areas of Santa Barbara County.

◆ In the town of Winters, housing developers were eliminating the natural habitat of hawks native to the area. Teens from the Woodland Community Continuation High School, as part of their community service project, built and installed hawk nesting boxes to replace the natural hawk nesting sites destroyed in Yolo County.

◆ In San Luis Obispo County, where erosion and sedimentation has been threatening plants and fish, a team of teens organized a coalition of environmental groups to present a "Watershed Awareness Day." Hands-on activities presented by the teens and younger participants helped inform community residents of the impact and dangers inherent in watershed degradation and demonstrated how local residents could get involved in restoration efforts.

◆ Eight teens from the Camp Tehema Migrant Education Program attended a 4-H SERIES training in Vallejo, then led 4-H SERIES activities during a five-day multicultural camp in Tehema County for 71 migrant youths.

◆ Twelve teen volunteers in Placer County targeted the issue of solid waste reduction by developing and implementing an education program and providing advantageous alternatives to landfills for residents of the Auburn community. Teens reached more than 23,000 residents with their message by setting up recycling displays and conducting informational activities such as local fairs, speaking to civic groups, and leading special programs in local schools.

◆ A Merced County teen designed a mobile 4-H SERIES Science Center that containing all the necessary materials to conduct 4-H SERIES activities at fairs, expositions, camps, museums, and schools. The Center included a computer so that participants could log onto the 4-HP state-wide electronic information system.

◆ Teen leaders from the Sonoma County 4-H Club and the Santa Rosa Economic Council collaborated in providing 4-H SERIES activities at a day camp for 140 youngsters from local shelters for homeless persons and battered women.

◆ Teen volunteers from the Martin Luther King Continuation High School in Yolo County collaborated with the Explorit Science Center in Davis by leading hands-on activities for elementary school groups visiting the Science Center.

The Impact of the Projects on Teens and Their Communities

Three sources of data were used to describe the impact of the projects on teens and their communities. These included portfolios developed by each team of teens to represent their individual project and its outcomes; four focus group interviews conducted with 28 teens as part of a teen symposium held near the conclusion of the project; and, transcripts of presentations made by six of the teens at a 4-H State Leadership Conference at UCLA in August 1995.

Evidence in the Portfolios

Teen leaders developed project portfolios for each of the 26 projects. All of the portfolios included a table of contents, goal statements, descriptions of evaluative procedures, photographs of project activities, and evidence of outcomes on the stated goals. Most included statements of unanticipated challenges that were recognized and overcome as well as examples of serendipitous, unanticipated benefits. The portfolios followed a report format ranging in length from 15 to 53 pages. Several included student artifacts such as videotapes, audio tapes and articles that had been constructed by teens and younger youths during project activities.

In each case, the portfolios documented evidence that the volunteer teen leaders and their younger participants were able to find and frame, and then to address an important issue facing their communities. They were also able to mount an initiative that, to some degree, mitigated the impact of one or more negative effects stemming from that issue within their communities. The evidence of "finding and framing" was found in the goal statements, the plan of work for the proposed project, and the summary or conclusion section of the portfolios. Evidence for the recruitment of younger youth, providing them with engaging learning activities, and enrolling them in the community-based projects was found for 25 of the projects among the artifacts included in portfolios (for example, attendance sheets, photos, and participant construc- tions from the actual activities). The single exception to this pattern was a project developed by a single teen. In this case, the teen built a 4-H SERIES Science Center trailer and, while the trailer could not be included in his portfolio, it constituted ample evidence of his accomplishments. In each of the portfolios the assistance and support of an adult coach was acknowledged.

Evidence from the Focus Group Interviews and Conference Presentations

The 4-HP teen leaders participated in focus group interviews as part of a symposium organized near the end of the 4-HP project. The focus groups were conducted by two experienced science educators. Each of four focus groups had approximately seven 4-HP teen leaders; two of the focus groups also included several adult volunteer coaches. The vignettes taken from the focus group interviews and the teen presentations at the leadership conference focused on leadership, teaching, cooperative work, community involvement, and sciencing.

Teens' Comments on Leadership

When teens were asked what they had learned about leadership from their experiences in the project, they commented:

◆ I think the biggest thing we learned out of it was how to be leaders. 'Cause a lot of us are kind of shy, and we weren't really into talking in front of people, and I think this really gave us self-esteem, to talk in front of people.

◆ Our group was teens generating solutions.

◆ It was fun. We just started doing it and then it just became a snowball.

◆ So what we did was we tried to make the community more aware of what we were doing...we had workshops.

◆ ...as far as the environment is concerned, mainly kids and teenagers are the ones who understand it...'cause we're the ones who are still going to be alive when all of this goes on...when it gets to the breaking point...we're the ones who have to figure out how to mend it.

Teens' Comments on Teaching

When asked about what they learned about teaching, teens responded with:

◆ I learned out of it [that conservation] is important, and if we don't teach anything to the kids it will never happen, and pretty soon the earth will be just a big landfill.

◆ The kids (in the Boys and Girls Clubs) loved the projects and that's the reason they wanted to sign up.

◆ Our district fair asked us if we would have a hands-on booth during the entire time the fair was open. People of all ages could come and do things that [had to do with] recycling, so they all make picture frames and puzzles out of all kinds of things...we had a large audience.

◆ ...if they get something out of it, that's the best thing you could do for them...it's the knowledge.

◆ The kids learn a lot about us [teens] and we learn a lot about them....you come in and try to educate them as well as yourself. Along the way you both get smarter.

◆ ...when it's hands-on...they like it...if they're sitting there listening to somebody talk it's totally different than being able to touch the snail's shell...and you know, how you can make your own paper...it's really cool!

Teens' Comments on Cooperative Work

When asked about cooperative work, teens commented:

◆ I think our 4-H SERIES team, the six of us, really learned how to work together as a team to teach the kids.

◆ We found that we did not have to have very much money to put on our program. Every place we went people said, "Oh, you can use this, that, or the other thing I have." So they were becoming aware of the potential value of the things that they no longer had use for and now we have all kinds of resources to use. We just have to be creative and find uses for them all.

◆ Their attitude changes from a "we can't make a difference because we're just one person and the world's doomed anyway" to an attitude of seeing the beauty around them, seeing what they can do to help, and pitching in to make this a better place to live.

◆ ...I worked with my friend Theresa, and we're both very similar and very different. And so, we got different input but we kind of, we don't argue or anything...you have a group of leaders, you have to make sure there's people who you can compromise with, they know how to work together.

◆ When you're in a group it's easier because you don't always have to do everything. You can do part of it, and then they'll do part of it.

Teens' Comments on Community Service and Community Involvement

When asked about these topics, teens said, for example:

◆ I think that what people in the community got out of it is that there isn't just one way of recycling...like just cans.... That there are different ways of recycling like composting with worms.

◆ This became a far more, far reaching [project] than we ever dreamed. It all started because the county saw the kids in action really making a difference in our garbage problem. They came to us to ask solutions to, you know, their problems.

◆ ...on a personal level...the kids went home and looked at where they lived and tried to make their home safer...

◆ We have found that, you know, we're not even considering stopping. We're just getting started with [tapping into] the interest in our community.

◆ ...a good project is when you go out into the community and you make them maybe see something they didn't see before...maybe something that was there, and you just kind of like opened a door for them.

Teens' Comments on Sciencing

Teens' views about sciencing were represented by statements like:

◆ ...in the snailing unit we were testing hypotheses...they'll say, "Well I wonder if this snail would do that if this happens" so you say, "Well, find out."

◆ ...they learn to ask those questions out loud and then set it up so they can figure out what the answer to that question is.

◆ I think the 4-H SERIES activities are really set up good so that once they start believing in themselves to ask the questions that they can be doing that throughout the program.

◆ We just used things around...not something that takes a scientific laboratory to do. It's just exploring the world around you. We found that everybody from the little kids on up to the adults had their eyes opened.

◆ I think this kind of activity [hands-on] is good for science 'cause it lets more people see science and see how there's more parts of science than just...just what you see in your science book.

Students involved in the 4-H SERIES "Getting to Know Worms" activity at the Placer County School Age Child Care Program in Auburn, California.

◆ You get back to the biology or whatever. Yeah, I learned that...and when I'm teaching...I'll completely forget that it's even science, and it's just fun.

◆ Science is everything. I mean it's not reading out of a book. Science is learning, and that's how it should be.

Discussion

The portfolios and interviews provide considerable evidence that science-related community service projects can help teens. The portfolios were especially effective in conveying a clear picture of teens making an authentic and substantial difference in their communities. After examining these sources, it seems clear that teens had many opportunities to use problem-identifying and problem-solving skills during their project activities. All of the projects involved allocation of resources and development of a time line. Teens inevitably had to generate explicit strategies for pursuing their project goals. While each of these circumstances created useful cognitive situations for teens, it is important to acknowledge that as teens went about finding, framing, and solving challenges they generally worked in social situations. That is, the projects involved working with a variety of persons and agencies. By making various agreements with other individuals and groups, for example by promising to complete a task by a certain time, teens made commitments to themselves and to others and thereby engaged a complex set of interdependencies. Making such arrangements and performing in the context of these interdependencies is precisely what leadership is all about. By initiating, designing and carrying out the projects, teens had many opportunities for exercising individual cognitive skills, but also they put themselves in positions that required various kinds of leadership. In this manner, teens acquired and exercised several of the core skills that are advocated for a successful work life (see for example, Reich, 1994; 1991a; 1991b).

There is a second aspect of participation in science-based community service projects that is important for some teens. The projects all involved opportunities for teens to engage in teaching science in out-of-school settings. These activities allowed those teens interested in pursuing a teaching career to get first-hand experiences with inquiry-based science instruction. Teens reported high levels of enthusiasm for using inquiry-based teaching strategies. And, "Inquiry into authentic questions generated from student experiences is the central strategy for teaching science," according to the National Science Education Standards (1996, p. 31).

Tying leadership opportunities to educational experiences and participation in community-based service projects encouraged teenage volunteers to practice leadership and volunteerism in real life settings. Teens had opportunities to consolidate science concepts and inquiry skills while developing and conducting their service projects. Having teenage volunteers engage in meaningful tasks, focusing on issues that they have identified as important to their community, recreates and reinforces their connection to the community, and prepares them for successful participation in the American workforce.

References

Berryman, S.E., and Bailey, T.R. (1992). *The double helix of education and the economy.* New York: Institute on Education and the Economy, Teachers College, Columbia University.

Erikson, Erik (1968). *Identity: Youth and crisis.* New York: W.W. Norton & Company.

National Research Council (1996). *National Science Education Standards.* Washington, DC: National Academy Press.

Reich, R.B. (1994). Getting your slice of the pie. *San Francisco Chronicle.* October 13, E-9.

Reich, R.B. (1991a). The REAL economy. *The Atlantic Monthly,* February, 35-52.

Reich, R.B. (1991b). *The work of nations: Preparing ourselves for 21st century capitalism.* New York: Alfred Knopf.

4-H SERIES Project. (1995). *4-HP final report.* Davis, CA: University of California at Davis.

Student carefully examines dragonfly during a 4-H SERIES presentation at the Jane Goodall Institute's Youth Summit at the Biosphere 2 near Tucson, Arizona.

Chapter 5

4-H SERIES as a Component of the South Carolina Teacher Cadet Program

By Charles Fisher & Janice Poda

Introduction

The state of South Carolina has taken national leadership in the development of programs for early identification and recruitment of prospective classroom teachers. Through a state-mandated state-wide arrangement, almost every high school in the state has a Teacher Cadet Program (TCP) that, in conjunction with various colleges and universities, introduces high school students to the rigors and rewards of a career in teaching. This large, well-established program has been operating since 1986.

During the 1993-94 school year, the TCP began to use 4-H SERIES materials and activities as part of its overall curriculum. This chapter explores the use of 4-H SERIES in this lighthouse program. The chapter begins with background material on the general development of *teaching academies* in American high schools. Then, after a brief description of the South Carolina TCP and its recent introduction of 4-H SERIES, we explore responses of cadets who have participated in 4-H SERIES training and teaching as part of their overall work in the TCP. The chapter concludes with a discussion of benefits and challenges associated with the introduction of 4-H SERIES from the perspectives of the TCP and 4-H SERIES respectively.

Background on High School Teaching Academies

During the second half of the 20th century, the supply of teachers for American schools has been uneven. Demographic, social, and economic factors have combined to make teaching an attractive career during some time periods and a less attractive one during others. Successive waves of urbanization have complicated the picture by placing ever increasing portions of the school-age population in cities and their surrounding suburbs. These forces, among others, have resulted in wide fluctuations in teacher supply, causing some content areas (for example, science and mathematics) and some regions of the country to experience chronic shortages of teachers.

To counter these forces, several states have developed programs to recruit talented individuals to teaching as a possible career. Most of these programs target recent college graduates who are not yet settled on a career path or professionals whose fields are oversupplied with job seekers who might take up teaching as a second career.

In the 1980s, a new type of recruitment program, designed for precollegiate candidates, began to appear in a number of areas of the country. These programs introduce secondary education students to teaching while they are still in high school and, in some cases, even earlier. Typically these programs, generically referred to as teaching academies, constitute a special option in the high school program and provide a variety of opportunities for participants to engage in classroom teaching activities. Such programs often have special arrangements with teacher training programs in local colleges and universities. The vast majority of teaching academies are relatively small in the sense that they operate in a single site with less than 50 prospective teachers per year.[1] The TCP in South Carolina is different from the majority of these teaching academies in scale, visibility, and method of support.

In the 1992-1993 school year, 4-H SERIES materials and activities were introduced into a typical teaching academy in California with considerable success. In the following year, 11 teaching academies formed a collaborative group known as the Science and Youth (SAY) project with the goal of introducing 4-H SERIES activities and materials into their already existing curricula. The SAY sites are located in large urban areas and, with the exception of the program in South Carolina, each involved a single high school. 4-H SERIES was especially attractive to these programs because, on the one hand, its content is science (recruiting and training science teachers had been a priority for some time—especially at the elementary school level) and, on the other hand, the pedagogy of 4-H SERIES is hands-on inquiry (a style of teaching and learning that has risen in popularity with the advent of constructivism as a contemporary framework for understanding human learning).

In the 1993-94 school year, most of the site coordinators in the SAY Project attended 4-H SERIES training sessions in their regions and began integrating 4-H SERIES as a strand in their local curricula. How the 4-H SERIES was implemented varied depending on the unique constituencies of each site. The remainder of this chapter focuses on the introduction of 4-H SERIES[2] in the TCP in South Carolina. As noted above, the situation in South

Carolina was very different from that of the other ten programs in the SAY project.

Description of the South Carolina TCP

South Carolina's ambitious and nationally-recognized TCP celebrated its tenth year of operation in 1996. The following excerpt from the 1995 annual program assessment completed by the South Carolina Educational Policy Center describes the overall program:

> The Teacher Cadet Program, sponsored by the South Carolina Center for Teacher Recruitment, is an innovative approach designed to attract talented young people to the teaching profession through a challenging year-long introduction to teaching. The program seeks to provide high school students insight into the nature of teaching, the problems of schooling, and the critical issues affecting the quality of education in America's schools. Piloted in 1986, the TCP grew to include 139 high schools serving 2,297 academically able high school juniors and seniors. To be eligible to participate in the TCP a student must have at least a "B" average, be enrolled in a college-prep curriculum, and be recommended by a least five teachers. During the year-long course cadets participate in seminars, group projects, and discussions with professionals in the field of education. Cadets study educational history, principles of learning, child development, and pedagogy. They visit classrooms to observe teachers and students, construct lesson plans, and experience some of the joys and challenges of being a teacher. Faculty members from 24 colleges and universities provide support for the 117 cadet sites participating in the college partnership. Depending upon the relationship between the high school and its college partner, cadets may receive college credit for their work in the TCP class. (Fanning and Ren, 1995, p. 1)

Since its inception, the TCP has been viewed positively both in South Carolina and by proponents of teacher recruitment programs throughout the United States. The program has undergone a yearly assessment conducted by the South Carolina Educational Policy Center. After the 1994-95 school year, the Center reported:

1. The TCP is attracting bright capable students who represent the top 20 percent of their high school classes and who score higher on the SAT than state and national averages.

2. The TCP is an effective tool for recruiting males and minorities into careers in education. There has been a 100 percent increase in the number of minority males choosing to become teachers at the end of the 1994-1995 school year.

3. The TCP is effectively encouraging teaching as a career choice with 36 percent of the cadets choosing teaching as a career.

4. Former cadets do become certified teachers in South Carolina as represented by 118 former cadets (of the 1989-1990 cohort) who are now certified to teach in South Carolina. This brings the total number of former cadets who are certified to teach in the state to over 700.

5. The TCP was instrumental in former cadets becoming teachers. Seventy-eight percent of former cadets who responded to the survey reported that they are currently teaching in South Carolina.

6. The TCP is working because of the effectiveness of the program components (the college partnership, the regional support model, and support of the South Carolina Center for Teacher Recruitment). (Fanning and Ren, 1995, p. 9)

Introduction of 4-H SERIES in South Carolina's TCP

In the 1993-94 school year, SAY/4-H SERIES was piloted in one high school in the TCP. That is, cadets received training in SAY/4-H SERIES procedures for teaching science and then used lessons from the 4-H SERIES program to engage youngsters in science activities in local elementary schools. The SAY/4-H SERIES activities became one plank in the larger ongoing program of the TCP. The following year, when the data for this chapter were collected, SAY/4-H SERIES was used as a portion of the TCP curriculum in six high schools. Use of SAY/4-H SERIES has continued to grow within the TCP. By 1995-1996, SAY/4-H SERIES was in use in 62 high schools and plans were in place to implement the program in the majority of high schools in the state.

To provide formative feedback to both the TCP and the SAY/4-H SERIES projects, a descriptive study of the implementation was undertaken. Visits to three of the six high schools using SAY/4-H SERIES materials during the 1994-95 school year were conducted in May. At each school, the site director was interviewed, a focus group was conducted with cadets who had participated in the program, and a team of cadets was observed while teaching in a local elementary school. In addition to these data, brief surveys about SAY/4-H SERIES were administered (both in the fall and spring) to all high school participants in five of the six SAY/4-H SERIES sites in South Carolina. These questions were included in the larger survey of TCP that is conducted annually by the South Carolina Educational Policy Center. Analyses of the data collected from these various sources are presented in the next section.

Selected Comments from High School Site Directors

In describing their programs, site directors consistently pointed out several features of the TCP that make it attractive to prospective teachers. Being admitted to the TCP carries considerable prestige. Participants in the program come from the highest quintile of students in the state in terms of academic performance. Once admitted to the TCP, students take coursework, observe in elementary and secondary school classes, and ultimately work in a field-based teaching assignment for six to eight weeks.

Site directors reported that SAY/4-H SERIES activities were offered as part of the program because they included inquiry teaching, hands-on activities, and science content. All three of these features were considered useful augmentations to the overall TCP curriculum. Although there may be some variation from site to site, SAY/4-H SERIES activities make up as much as 20 percent of the total program. After SAY/4-H SERIES training, cadets have the opportunity to teach a 4-H SERIES lesson in four to five different elementary school classrooms. The SAY/4-H SERIES fieldwork is independent of the internship assignment but, in some cases, cadets might make use of SAY/4-H SERIES activities in their internship classes.

Site directors were asked what benefits cadets get from participating in the SAY/4-H SERIES portion of the program. One veteran TCP teacher put it this way:

> One of the problems with education, I think, is that so much of it is cut-and-dried lecture work, especially at their level [seniors in high school]. And I think, if nothing else, it [SAY/4-H SERIES experiences] gives these college-oriented high school students a chance at looking at another type of teaching, instead of...just lecturing. We need to do a whole lot more hands-on inquiry-based instruction from the ground on up. We're not doing enough of that, especially at the secondary education level. And I hope that by stressing SAY/4-H SERIES so much, they will keep in mind that there is another way to do things. When cadets go back into a classroom, [they might say to themselves] here's my content material, now let me think of some way that I can turn this into an inquiry-based lesson with process skills and activities, that type of thing.

The site directors were quick to point out that the process of becoming an accomplished teacher is measured in years. In the context of an introduction to teaching, SAY experiences are intended to "get the attention of cadets."

> These young people grasp the fact that there's a whole lot out there that they don't know and sometimes they need to search it out for themselves. That's what the [SAY activities] are supposed to do—get their interest and have their [elementary school] children start asking questions—not so much to answer all their questions, but to require them to use their senses and their thinking skills to try to think of answers themselves, or find out how to find out the answers for themselves.

While hands-on learning activities, similar to those used in SAY/4-H SERIES, have been advocated for decades, they have not transformed pedagogy in American schools. One of the major barriers to the adoption of hands-on learning in the public school classroom is the constant need to replenish consumable materials. This time consuming and sometimes expensive task has eventually overcome early enthusiasm for the method in some cases. Since the everyday materials used in SAY/4-H SERIES lessons need continual restocking, we queried site directors about this issue. One site director reported:

> But as far as purchasing the materials and finding the things, it's been real easy. We've been able to use a lot of things we find at home, and my cadets are real good at going through their list and going, "Okay, I've got this, I can bring this. I've got some of this at home. No, problem, my mom collects these," and that type thing. So they're real good at helping me by bringing in their own materials as much as they can. It gives them some investment in it, too.

Because SAY/4-H SERIES is being introduced into an established program, we wondered if there might not be some special start-up problems. When asked about such problems, site directors mentioned logistical concerns about having teams of cadets in the field relatively often. However, it seems that most of these concerns had been alleviated in the first few months of operation. An equally predictable concern arose around the changes in the TCP curriculum caused by the introduction of SAY/4-H SERIES. One site director commented:

> The biggest problem is that I'm behind in my curriculum, my teacher cadet curriculum, because SAY/4-H SERIES is taking more time. It's one more thing to put in and it takes so much time out of the actual classroom. Those six or eight different outings, and I'm sure if I counted it up it would be a whole lot more days than six or eight, those were days that were set aside in my curriculum for the other activities. Now, I feel like it's been a [productive] trade off. I have not had a problem with that, because the things my students are learning in SAY/4-H SERIES and the experiences they are having outweigh some of the things that I would have put into the curriculum instead of that.

Cadets' Responses to SAY/4-H SERIES

In the 1994-95 school year, of the six high schools using SAY/4-H SERIES as a part of the TCP curriculum, cadets in five schools were surveyed in the fall and spring as part of the larger annual TCP assessment survey. At the time, one high school was in its second year of SAY/4-H SERIES implementation and the other five schools were in their initial year. The TCP had been operating for about a decade at all six school sites. Responses to the survey provide an overall indication of how participants viewed SAY/4-H SERIES and, to a great extent, science teaching in the context of their TCP experience.

Machine-Scored Item Responses

Cadets responded to ten statements on a five-point Likert scale, ranging from strongly agree (5) to strongly disagree (1). Responses were electronically scanned and averages for each item were calculated. The survey questions and average responses are presented in Table 1.

From the first two items in Table 1, note that students' endorsements of science as a favorite subject rose from 3.1 in the fall (before participation in SAY/4-H SERIES activities) to 3.4 (after participation in SAY/4-H SERIES activities). The next two items indicate an increase in interest in teaching science when responses in spring were compared with fall responses. On average, cadets were slightly negative about the possibility of teaching science before engaging in the program and positive about it afterward. Cadets tended to attribute this change to experiences in SAY/4-H SERIES (see item 5). Items 6 and 7 indicate that cadets took more than the minimum number of courses in science in high school and that they intended to continue this trend in college. Items 8, 9, and 10 indicate that cadets strongly endorse the use of hands-on science learning in elementary schools.

Open-Ended Survey Responses

In addition to the 10 machine-scored items included in Table 1, cadets also responded to several open-ended items as part of the spring survey. One of these questions asked cadets what they had learned about teaching as part of

Table 1
Cadets' Responses to Selected Survey Items

Occasion	Survey question	N	Average
Fall 94	1. Science is my favorite subject.	72	3.1
Spring 95	2. Science is my favorite subject.	69	3.4
Fall 94	3. If I were a teacher, I would want to teach science.	72	2.7
Spring 95	4. After participating in the SAY Program, my interest in teaching has increased.	69	3.3
Spring 95	5. After participating in the SAY Program, my interest in science has increased.	69	3.4
Fall 94	6. In high school, I have taken only those science courses that are required to graduate.	72	1.9
Spring 95	7. In college, I will take only those science courses that are required to graduate.	69	2.8
Fall 94	8. In elementary school, students do learn more about science from "hands-on" activities.	72	4.4
Fall 94	9. In elementary school, students do learn more about science from listening to the teacher and reading the textbook.	72	2.3
Spring 95	10. In elementary school, students learn more about science from "hands-on" activities than from listening to the teacher and reading the textbook.	69	4.7

their experience in SAY/4-H SERIES. Statements from the 62 respondents contained 82 brief accounts of learning. Cadets responded most frequently with a statement about the efficacy of hands-on activities with elementary school students. This response occurred 27 times representing 33 percent of the overall responses. The next most frequent response (15 occasions or 18 percent) referred to classroom management issues and concerns about getting and maintaining students' attention. It is interesting to note that concern about classroom management is traditionally the primary concern of novice teachers; in this case, that concern, though still very important, was displaced by the perceived efficacy of hands-on activities as a central component in elementary science teaching and learning.

Ten of the cadets' responses (12 percent) noted that teaching was harder than they had expected it to be. They noted that you had to be prepared and that teaching was time consuming well beyond the contact time with students. Three responses—that teaching required patience, that teaching can be fun, and that teachers must be flexible—each garnered an additional 6 percent of the cadets' responses. Cadets mentioned 11 other "things learned" but none of these other responses was mentioned more than twice.

Focus Group Responses

In early May, focus groups of six to ten cadets were conducted in three participating high schools to discuss the SAY/4-H SERIES teaching experiences. The hour-long meetings took place in their respective high schools. Each meeting was tape recorded.

As part of the focus group discussion, cadets commented on what they had learned about teaching and about themselves during their SAY/4-H SERIES experience. Their responses were similar to those obtained from the year-end surveys. In two of the three groups, there were repeated expressions of enthusiasm for hands-on science learning. Cadets reported that both they and their younger students liked to do the activities:

◆ Yeah they loved the hands-on part...when we came and did "It Came From Planted Earth," we did worms and they liked the worms. In most classrooms, you know, they talk about animals and they just say this and that and you tell them what the animal is and what it does. But they never really, you know, like just go outside in the yard and see the animal you are talking about in class...and so they can grasp it, the hands-on part, and they loved that. I loved that, too, hands-on. It really gets them excited about it when you can connect it to what they've been doing and what they're going through.

◆ The movement involved in the demonstrations definitely kept their attention better than just lecturing to them or that type of teaching, and, so, I'd say it was very effective in that respect. They really like the hands-on stuff.

During the focus groups, several participants seemed to equate hands-on science activities with inquiry learning indicating that their notions of inquiry learning are still developing. In one of the focus groups, some cadets expressed frustrations with the open-ended lesson structures. They felt that their students wanted clear answers to their questions and they were somewhat uncomfortable with either not knowing the answer themselves or encouraging students to struggle a little more. Although these concerns did not surface in the other groups, this may be a healthy sign since learning to teach through the inquiry method is challenging and the circumstances in which cadets most often work (single class periods with a given group of students) are not particularly hospitable to extended inquiry.

Cadets reported new learning in the general area of classroom management. The focus group participants, echoing the survey results, saw the need to be flexible in the classroom. Some cadets expressed it this way:

◆ You can't go in every day with something absolute because you never know what's going to come up during that class period.

◆ Sometimes we had kids construct things. In the earthquakes lesson, we would do buildings using clay for ground and toothpicks for making the buildings. When it comes to...construction...it can drag on forever. You can't anticipate how long it's going to take.

Teen leaders are involved in sorting and classifying as part of their training in use of science processes.

♦ So, it's kind of unpredictable then, in terms of time; you certainly don't know what's going to happen each time.

♦ You have to learn to...think on your feet and answer some of the most off-the-wall questions that do come up.

♦ You have to keep things moving all the time. The kids get antsy if you, like, get slow about things.

Some cadets also commented on disciplining students, a perennial concern for both novice and experienced teachers:

♦ Walk softly and carry a big stick [laughter]. We had a time when we were doing "Beyond Duck and Cover" when we looked in the unit and there's a lesson that has a tin plate with Jell-O and clay. And no matter how hard we tried...in fact, we kind of got into it one morning because [the students] asked if they could start eating, and so I'm like, "Sure." And then as teachers, you know, we decided that they weren't going to eat. But it's still hard to tell them, "No, you can't eat anything. Don't pop those marshmallows."

While cadets grappled directly with such issues as classroom management, they also gained insight into some of the central issues of teaching. For example, there was evidence that cadets became aware of the perspective that is necessary to allow teachers to see students differently from themselves and differently from other students. As cadets decribed it:

♦ It sort of makes you realize that you can't take for granted that everybody knows what you do.

♦ Making it interesting, and relating it to them. Everything has to be relevant.... Being able to present...information in a way they can understand it and still keeping it challenging and keeping it interesting and fun. But not being boring by presenting it in a way maybe below their understanding, but not making it too difficult for them to understand....That's a real challenge.

Cadets reported that teaching is more complicated than it sometimes appears. There are many things that impinge on how a particular lesson goes. In most cases, cadets acquired new respect for their own high school teachers:

♦ You have to deal with a lot of outside pressures, and, I mean...it would have gotten me off track. It's a juggling act definitely and it gave me respect for that juggling act that I didn't have before.

♦ Sure, I mean, definitely I have more insight into what kind of preparation they [teachers] have to go through and what kind of work they have to put into it to be able to present the information in a way that is fun and challenging; to get it at the right level for each of the students; and so I definitely have more respect for all my teachers.

Cadets' Suggestions for Improvement

As part of the spring survey in 1995, cadets were asked to suggest things that would improve the SAY/4-H SERIES portion of the TCP. The cadets generated 64 suggestions. The most frequent response concerned the 4-H SERIES curriculum. Requests for more variety in the curricula, more activities to choose from, more activities that are adaptable to a range of grade levels, and more interesting activities accounted for 13 suggestions (20 percent). Interestingly, the second most frequent response, accounting for 10 suggestions (16 percent), stated that no changes were needed. In these comments, cadets were often explicit about the fact that they liked the program very much the way it is. The third and fourth most frequent responses (9 suggestions, 14 percent each) focused on the need for more training on the 4-H SERIES curricula. Most students would like to have had two weekend training sessions so they could be fully trained in additional 4-H SERIES units.

Nine cadets expressed an interest in having more opportunities to teach the 4-H SERIES lessons in elementary schools. A few of the suggestions in this latter category called for efforts to increase the number of elementary schools participating in the program. Two other suggestions made by several of the participants (4 percent of the total) called for more time with a given class of elementary school students (so that projects could be completed), and development of a funding arrangement that would provide small amounts of cash to replenish SAY/4-H SERIES materials boxes after each use.

Discussion: The 4-H SERIES Perspective

Generally speaking, bringing SAY/4-H SERIES into the TCP appears to be a productive marriage with plenty of potential for growth. From the data available for this study several things seem clear. First, SAY/4-H SERIES was introduced into several TCP sites without major problems. Although secondary school teaching academies were not among the contexts for which 4-H SERIES was originally designed, TCP site directors were positive about the program. They found the content and pedagogy of 4-H SERIES to be compatible with their existing programs and were able to manage the logistics associated with the science materials and field placements without difficulty. The ease with which this introduction has proceeded may be due, in part, to the fact that the TCP is a mature program with established goals and procedures and stable competent leadership. The incorporation of SAY/4-H SERIES was undertaken as an intentional change in one part of the TCP program.

Second, SAY/4-H SERIES was well received by cadets. Positive shifts in their attitudes toward science and science teaching were correlated with experience in SAY/4-H SERIES activities. Cadets were especially enthusiastic about use of hands-on science activities in elementary schools and they reported gaining important insights into teaching from their experiences. While this latter point is undoubtedly a result of both the overall TCP as well as the SAY/4-H SERIES component, most cadets had praise for the science-oriented work in schools.

Third, SAY/4-H SERIES activities apparently were easily integrated into the more structured context of formal education. Having been designed for use in informal settings, the training, content and learning activities were flexible enough to be carried out in regular classroom environments.

Most cadets were satisfied with SAY/4-H SERIES lessons even though the contexts in which they used them differed distinctly in terms of two of the primary 4-H SERIES design conditions:

◆ One, the activities were originally designed to be taught in a series of meetings. Cadets generally taught the lessons as single-occasion units. That is, teams of cadets would go to an elementary school classroom on one occasion to work with a given group of children and not return for follow up. One implication of the single-occasion meetings is the lack of time for children to engage in extended inquiry. Some cadets, though a very small number, touched on this point when they recommended more time with a particular group of children as one way to improve the SAY/4-H SERIES component of the program. As long as SAY/4-H SERIES remains organizationally separate from the regular TCP field placement, it is difficult to see how this situation could be substantially changed. Perhaps some cadets could choose to use SAY/4-H SERIES activities as a portion (say one week) of their larger field placement. (The TCP, however, is an introduction to teaching, not an introduction to science teaching.) One possible disadvantage of maintaining the single-session format is that cadets may not get a useful experience in leading and facilitating inquiry learning, even though this is one of the rationales for incorporating SAY/4-H SERIES activities in the TCP.

◆ Two, the culmination of a curriculum unit in a community service activity related to the content of that unit is generally missing in the TCP context (at least in this early stage of the introduction of 4-H SERIES into the TCP). This feature, while theoretically attractive, has been difficult to implement consistently in informal education settings and its absence in school settings is not surprising.

Notes

1. See the following chapter for a description of 4-H SERIES in the context of a single school teaching academy.

2. In this report, we use the combined acronym SAY/4-H SERIES to remind readers that 4-H SERIES activities, materials, and training are being used in the context of existing high school teaching academies.

Reference

Fanning, M., and Ren, J. (1995). *Assessing the teacher cadet program: The 1995 study.* Columbia, SC: The South Carolina Educational Policy Center.

Teens share their 4-H SERIES "It Came from Planted Earth" project with the community through a display at the Exploratorium in San Francisco.

Through a program in which scientists and engineers from the community mentor teens, employees of the IBM Corporation introduce students to basic chemistry as part of the Indendence High School Teaching Academy in San Jose, California. The next step will be for the teens to teach hands-on chemistry activities to elementary school students.

Chapter 6

4-H SERIES
as an Educational Catalyst:
The Evolution of Project Excel

By Charles Fisher

Project Excel is a complex multi-agency science learning collaboration in Santa Clara County (California). There are several direct ties between Project Excel and 4-H SERIES, making the development of Project Excel particularly instructive in terms of 4-H SERIES as a tool for strengthening science learning among children in both formal and informal settings. While there were multiple connections between Project Excel and 4-H SERIES, these connections changed substantially during the development of the Project. This change is one part, but by no means the only interesting part, of the story.

During the school year immediately before the inception of Project Excel, the high school and elementary schools that would subsequently constitute a primary target audience for Project Excel were already engaged in 4-H SERIES activities. These schools were the pilot sites for using 4-H SERIES materials and activities in the context of a high school teaching academy. This innovative use of 4-H SERIES came about in part through the leadership of the director of the Teaching Academy at Independence High School in San Jose, California. This pilot experience contributed significantly to the conceptualization of the Science and Youth (SAY) project.[1] Given this beginning, Project Excel initially looked like a typical one-site implementation of the SAY project in its first year. However, by the end of the third year, the goals of Project Excel had shifted and were strongly focused on community development.

This chapter describes some of the key processes and outcomes of Project Excel as it unfolded (a process that is still ongoing). We begin with a description of the original project followed by data collected annually as part of efforts to assess the impact of Project Excel. The conclusion of the chapter provides a discussion of the Project's development and suggests some tentative interpretations of the data.

An Introduction to Project Excel

Objectives and Major Components

Project Excel began as a youth development program carried out by the University of California Cooperative Extension in Santa Clara County in cooperation with Eastside Union High School District, Alum Rock School District, and other agencies in Santa Clara County. At its inception in 1992, Project Excel had the following outcome and process objectives:

◆ 9-to-12-year-old children will participate in an increased quantity and quality of science experiences.

◆ Teen leaders and 9-to-12-year-old children will increase positive attitudes toward science.

◆ 9-to-12-year-old children will demonstrate positive leadership roles and initiative in applying science to home and community service projects.

◆ Project Excel will significantly increase parent and community volunteer work with elementary children leading to new nonformal youth/science program activity in elementary school neighborhood. (Project Excel Quarterly Report, November 1992)

As time passed, the last of these objectives grew in importance compared to the others.

As originally conceived, Project Excel had two major components: (1) The in-school or cross-age teaching component; and (2) The out-of-school or community service and development component. The first component was implemented by students participating in Independence High School's Teaching Academy, a careers-in-training exploration program. The teens, after training in the use of 4-H SERIES curricula, had the opportunity to conduct 4-H SERIES inquiry learning activities with elementary school students in structured classroom lessons and community action projects. Volunteer scientists served as coaches and role models for the teens as they learned teaching and leadership skills.

The second major thrust of the program engaged teens, younger youth, and neighborhood residents in a variety of community service and community development projects (the out-of-school component of the program). As the project has evolved, the role of the out-of-school program has increased in importance. In Year I, the out-of-school program focused on community awareness issues and was relatively separate from the cross-age teaching component which, at that time, was the central part of the overall program. By Year III, Project Excel had established its focus on school communities and based its program on after-school activity programs for elementary school students, parent education programs, family science days, science camps, and leadership programs as well as cross-age science instruction. As part of this refocusing, the Department of Recreation, Parks, and Community Services of the City of San Jose was added to the partnership.

A large majority of participants in the program, both high school and elementary students, live in conditions of poverty and, for many, English is a second language. Historically, these students have been at risk of failure in school especially in areas involving mathematics, science, and technology.

Project Excel's First Year

Based on an examination of a wide variety of Project Excel documents and on data collected during a one-week site visit, Project Excel was described as an ambitious, complex and thriving project. During this start-up phase, Project Excel had successfully begun operation and a variety of mechanisms had been developed to support it. In spite of the complexity of the collaboration (two school districts, University of California Cooperative Extension, the United States Department of Agriculture, several scientific and engineering professional agencies, and several community agencies), bureaucratic delays during the first several months of funding, and the fact that the project leaders also had time-consuming commitments to other high-priority activities, Project Excel made remarkable progress in its first year. While there was progress in every area of the project, there was more in some areas than in others.

From examination of project documents and data collected during the site visit several aspects of the project were identified as strengths. The cross-age tutoring as implemented in the project appeared to engage both teen leaders and elementary school age youths. Teachers reported lower rates of absenteeism among their students when teen leaders were scheduled to work with their classes and both teens and elementary school students reported that the cross-age tutoring was challenging and exciting. A second aspect of Project Excel that was often mentioned as a strength was the organization of teens into teams to work on and implement hands-on science experiences for youths. Teens reported learning from each other and being supported by other team members both when work was going well and especially when individual team members either lost focus or motivation to complete their contributions to the team's work.

The curriculum tasks and activities suggested in the 4-H SERIES curricula were very well received by elementary school students. Active engagement with hands-on tasks apparently had strong appeal for students, regardless of setting. Preliminary concerns about the appropriateness of 4-H SERIES tasks for elementary school children came from two sources. First, would the tasks be interesting and engaging for the students, engaging enough so that elementary school students would cooperate in acting on them? Second, would the tasks fit the district and state science curriculum sufficiently well to garner support from school sites? While most elementary school teachers viewed 4-H SERIES tasks to be appropriate for their students, they also suggested that units were shorter than they would typically design themselves for use in school classrooms. These responses seemed to be motivated in part by the teachers' desire to extend the activities into additional sessions with their students between the occasions when teens worked in their classes.

Project Excel identified and recruited scientist volunteers to mentor teens in their work with elementary school students. The project coordinators are to be commended for establishing viable links between local scientists and participating youths. In some cases, exemplary mentoring occurred. In spite of these successes during the first year, there was considerable variability in the manner that scientist volunteers participated in Project Excel.

Volunteers came primarily from two sources: those recruited through contacts in the Society of Women Engineers (and other professional organizations); and those recruited from the Alameda Research Institute. Although an orientation to the project and opportunities for the scientist volunteers to meet and talk with teen leaders took place, only a relatively small number of volunteers were able to attend. As a result, some volunteers began working with teams of teens after having training in the 4-H SERIES content and pedagogy and an orientation on their roles in the project (training on mentoring and coaching) while others had none of these prior experiences. For those volunteer scientists who did not receive the orientation, several assumed that their role was to teach teens about science itself. In several instances this strategy was not productive and left the volunteers somewhat disturbed about their experiences with teens. In other cases, volunteer scientists who cast their roles more as coaches and general support people (and less as content experts) were generally held in high regard by the teens and reported high levels of satisfaction themselves.

The greatest single area of need for Project Excel appeared to be community liaison. While the project was very active in this arena and produced some high profile events, these activities were not well integrated with the overall goals of Project Excel. Staff of the elementary school, who were generally very supportive of the project, reported inadequate lead times for community meetings. In addition, parents of students reported disappointment and "misunderstandings" regarding changes in summer opportunities for their children.

A central design element of the program involved providing youth with practical opportunities to apply their scientific knowledge to community action projects. While the youth in Project Excel did engage in community action projects, these projects in the main did not relate directly to what they were learning in their classroom experiences.

Project Excel is intended to affect the attitudes of youth and teens toward science and scientific careers. From their enthusiasm for the project expressed by elementary school youth, there can be little doubt this was accomplished. In terms of attitude measurement, 48 elementary students from participating and non-participating classes were interviewed at some length. After analyses, there appeared to be no substantial differences between the attitudes of participants and non-participants. However, when attitudes toward science were examined by gender, girls who participated in Project Excel had substantially more positive attitudes than girls who did not participate.

In summary, strong progress was made in most areas during the project's first year of operation. The goals for Year I were either met or closely approached. The project was viewed positively by participating teens, youth, elementary school teachers, parents, and most volunteer scientists. At the end of the first year, it was determined that adjustments and changes needed to be made in the following areas: (1) the community liaison function; (2) the use of scientist volunteers; (3) the redirection of the community service segment of the curriculum; and (4) an earlier startup in the school year for teens and elementary students.

Project Excel's Second Year

In its second year of operation, Project Excel, to a great extent, sustained its general direction and intensity. Elementary school students continued to enjoy the hands-on science activities, had positive attitudes toward science and scientists (especially females), and increasingly participated in community development projects in their neighborhoods and beyond. Teens participating in the project received a substantial introduction to classroom teaching. They gained important insights, sometimes hard won, about the joys and tribulations of the profession, about working in teams in situations that require social responsibility, and about themselves as learners and teachers. A very positive aspect of the project for teens was the number and variety of leadership development opportunities that were made available to them. In Year II, teens' levels of participation in project outreach activities increased.

Project Excel continued to garner and maintain high levels of commitment from the agencies that cooperatively ran the program and the individual people who put in hundreds of hours over the course of the year to implement it. In addition to staff members of the Division of Cooperative Extension of the University of California, Independence High School, and the Alum Rock School District, Project Excel continued to attract an active group of highly qualified volunteer scientists and engineers who supported the program. Several parents of elementary school students in the program were contacted and they gave clear, strong and positive support for both the effects of the program on their children and for development of outreach activities to involve children, parents and other community members in community service activities.

A distinct highlight in Year II was the success of its community development program. This portion of the program was redesigned and strengthened after the initial year of the project. By the end of Year II, the after-school and weekend activities were providing both family-oriented and professional development opportunities for an increasing number of teens and elementary school-age children. A second annual community day drew approximately 800 people; local community activities like bread-making and kite-flying were getting students and parents together during out-of-school times; a parent university was successfully piloted at one school; and teens participated in a variety of leadership development activities by acting as counselors at camps, teaching hands-on science both in formal and informal settings, and touring and presenting at local science and engineering companies. Some teens traveled to Pittsburgh and Chicago to make formal presentations at national conventions on their experiences in the project. These and other activities underscored the strength of the community development and youth leadership development aspects of the project. Practically every group that was interviewed about the project, especially parents of participating youths, commented positively on the new directions that the out-of-school program had taken. This success was, in part, attributed to the hard and effective work of the university-based staff.

The cross-age teaching portion of the program operated in 16 classrooms in four elementary schools. In Year I, this part of Project Excel was the primary strength of the program. The in-school program continued to provide benefits for teens and elementary school students in Year II even though the community development segment played a stronger role in the program. The teens generally found the work difficult in the elementary classrooms but were vocal in their positive assessment of their experiences. Elementary school teachers were also generally positive about the program. While there were many positive statements about the Year II cross-age teaching program, there were indications that teens were sometimes not as well prepared as they had been in Year I, that materials were sometimes missing; that the cooperative learning teams had some difficulties; that the curriculum was not sufficiently integrated with the new district science curriculum; and that the scientist volunteers were not being given clearly defined roles.

There was one extenuating circumstance that affected the cross-age teaching program adversely. Both school districts, reflecting the general state of public education in California at the time, were beset with fiscal and other constraints (especially the elementary school district). "Hard times" made working conditions for teachers less than attractive and affected morale negatively. While this was not the only factor involved, the teachers participating in the Project, especially the staff of the Teaching Academy, were highly stressed during the year. They had less time (in most cases, no time) to observe teens teaching in the elementary schools and little time to debrief teens about their teaching performances or other aspects of teaching without compromising other parts of their jobs. Regardless of the source of the difficulties and the fact that resource cutbacks were not likely to change in the short run, several aspects of the program were identified for redirection in Year III.

In order to improve the quality of the teaching experiences for teens and, at the same time stay within the resources available, the cross-age

Two elementary school students demonstrate a chemistry module on "Jaime Escalante Community Day."

teaching portion of the program needed to be reduced in size. Without changing the number of teens involved in the program, the plan was to focus 4-H SERIES activities in two elementary schools. The scientist volunteers' roles and expectations constituted a second area requiring attention. In spite of clear signs of problems in this area during the first year, apparently there were not appropriate resources to work on this area and it continued to falter. These areas notwithstanding, the project involved collaboration among several long-standing public and private entities and represented a new way of doing business for each of the cooperating agencies. Given the kinds of changes that Project Excel was attempting to foster, it may take several years to get most aspects of the program appropriately refined and at least several years before any kind of institutionalization of the changes can be expected to occur. With this perspective in mind, Year I of Project Excel could be represented as a fast start, with the cross-age teaching component leading the way. In Year II, major improvements were made to the community development program while the cross-age teaching program received somewhat less attention.

Project Excel's Third Year

In Year III, Project Excel continued to expand with considerable success. The project had been evolving toward a clear focus on community development. In Year III, this focus was further established by creating and implementing after-school programs on an everyday basis. Maintaining participation of its new partner, the City of San Jose's Department of Recreation, Parks, and Community Services, was a necessary step in strengthening the elements of the community development program. Restricting attention to two elementary schools and their communities also made the after-school programs more manageable. The growing edge of the project was now clearly located in the community development component.

While activities centered on the elementary schools expanded, the cross-age teaching component of the program diminished somewhat but was still a viable if not dynamic segment of Project Excel. Expectations for the cross-age teaching component were reduced accordingly. Several structural elements were modified during the year: for example, teen teaching teams stayed in one classroom for a longer period of time; and the number of curricula to be implemented were reduced. High school teens continued to find the cross-age teaching challenging but rewarding.

Project Excel collaborators expanded and introduced innovations in the community development arena while maintaining a viable cross-age teaching program. In this context, the project was perceived very positively by most of its stakeholders and clients. Elementary schools students, now the primary target audience, continue to be positive about the program; elementary school teachers and one of the school administrators are strong proponents of Excel. Parents in the two elementary school communities value the project highly and continue to request more services.

Against this generally positive background, project collaborators identified two areas of concern at the end of its third year. One of these areas was all too familiar. The participation of volunteer scientists has declined considerably. Seeds of this decline had been apparent almost from the beginning of the project. There are several likely reasons for this decline. A prominent reason is that an appropriate role for volunteer scientists was not successfully developed. It would appear that the scientist mentor role is not viable in this project. It is not primarily science that seems to be attracting teens and when scientists volunteers have been successful it has been on the basis of their individual abilities to "hang out" with teens and provide social support in terms of organization and communication skills more than on their scientific knowledge and skill. If this assessment is accurate, we may need to rethink the current role for volunteers. Several options were considered. On the one hand, the volunteer scientist component could be terminated with relatively little effect on the overall program. On the other hand, the volunteer effort could be redirected to other areas of the project. For example, the after-school program could easily use several adults to help with its operation. A new role for volunteers could be defined in this arena, however that would also imply identifying a new pool of volunteers. These two responses represent extremes and, by the end of the third year, we were considering several intermediate alternatives as well.

The second area of concern arose from the successive bifurcation of the overall project. As the project has moved affirmatively to focus on community development and the provision of multiple opportunities for elementary school-age children to develop their academic and social skills, there has been a concomitant reduction in the overall role of the cross-age teaching component. By the end of the third year, the community development activities and the cross-age teaching activities were operating relatively independently. There was no apparent conflict between the two arms of the project, but no synergy either.

Project Excel as Viewed by Participants and Clients

While the previous section gives an overview of the initial years of Project Excel, it may be useful to explore the responses of the staff and clients to their experiences during Project Excel activities. At various times during the first three years of the project, data were collected via interviews with participants, paper and pencil responses by elementary school students and tenth graders, and portfolios constructed by high school teen leaders. Data on attitudes toward science were collected by Project Excel staff or teen leaders and the interviews were completed during a week-long site visit in the spring of each year. Data presented in this section are excerpted primarily from the third-year assessment of Project Excel (Fisher, 1995; see also Fisher, 1996, 1994, and 1993).

Interview Data

Teen leaders in Project Excel (sophomores) were interviewed in focus groups of approximately one hour each. Members of the Project Excel staff (including staff of the Independence High School Teaching Academy), volunteer scientists, elementary school administrators, and members of the after-school staff were interviewed individually. In addition, focus groups for teachers at participating elementary schools as well as groups of parents of elementary school participants were conducted. In each case, interviews and focus groups elicited descriptions of program operations and perceptions of strengths and weaknesses of the project.

Survey Data

Excel participants in the Teaching Academy responded to a spring survey after they had completed one year of participation in the project (and two years in the Teaching Academy). The survey elicited information about their perceptions of Project Excel, teaching, and themselves as learners.

Data on Attitudes Toward Science

Data on attitudes toward science were collected from third, fourth, and fifth graders. Project Excel focused its work in several elementary schools that were part of the "feeder" schools to Independence High School. Data were collected using two approaches to attitude assessment: individual students were given a structured interview (an adaptation of the Projective Test of Attitudes (PTOA) (Lowery, 1966, 1967), and the Draw-a-Scientist Test (DAST) (Chambers, 1983; Mason, Kahle and Gardner, 1991).

Project Excel Portfolios

Each teen leader in the Excel Program developed a portfolio documenting his/her experiences in the program and providing written reflections on those experiences.

Project Excel as Viewed by Participating Adults

Project Excel from the Perspective of the University Staff

When asked about the most recent accomplishments of the project, university staff responded first by describing a shift in focus to providing services to school communities. This focus had been developing since the inception of the project but in Year III, the staff was intentionally reorienting its efforts in this direction. The thrust of this idea is to develop multiple services within a community and thereby increase the probability of having an impact on a substantial number of children. This more systemic approach led to increases in the number and kinds of after-school programs in which elementary students could participate. As part of this shift from an emphasis in Year I on in-school cross-age tutoring to implementation of several educational and leadership development programs for children in Year III, Project Excel restricted its focus to two elementary school communities in the Alum Rock School District (down from four schools in the earlier years). A new arrangement with the City of San Jose's Department of Recreation, Parks, and Community Services allowed each target community to have an after-school program for elementary school students.

As a result of this shift, the university staff was heavily involved in orchestrating multiple programs and building and maintaining an infrastructure to support the programs. The contexts of the two communities, centering on Goss and Cassell Elementary Schools, were seen as important mediators in this process. For example,

> Although they are very close in proximity, they're very, very different [in terms of] community involvement and [distribution of] first, second, and third generation people that have been in the community. Especially at Goss. Goss is more a migratory community, and it's been very difficult to establish long term programs there partly because it's difficult to identify key leaders in that community. Since the population is highly migratory, any key leader may be gone in six months, or a year. Then they have to start the process over again. But at Cassell, second and third generation community members have been there, went to high school there, their kids went to high school there, have grandchildren in school. In the [Cassell] community you have some who have been there for 20, 30 years. And so key leaders are stable in the community. You also find that a lot of people...are really bilingual, whereas in the other one, it's primarily Spanish speaking, with very limited English. There are also differences in level of education and in familiarity with how the education system works.

After-school programs at both elementary schools were operating with good attendance.

Both after-school programs were supported in part by high school volunteers from the Independence High School 4-H Council that Project Excel had developed in the previous year. As a result, children at both school sites received science lessons from teens as part of their in-school and after-school activities. However, the vast majority of learning activities in the after-school programs came from other sources than 4-H SERIES and, although many activities dealt with science topics, a broad array of other topics were also introduced. After-school activities were voluntary for the children and, though there was considerable overlap in groups participating in in-school and after-school activities, the after-school groups drew from a much larger population. Both in-school and after-school programs were viewed by the staff as supporting the same learning goals for participants.

After some initial delays and misunderstandings in the new relationship between Project Excel and the Department of Parks and Recreation, that partnership "has gotten stronger as the months have gone by." By the end of Year III, the university-based staff viewed the after school programs as an indispensable part of the project. Since the Department of Parks and Recreation runs separate after-school drop-in programs at each of the schools, there has been a strong attempt to develop and maintain a distinct identity for the Project Excel after-school programs. The key issues here are that the Department of Parks and Recreation programs are intended to be drop-in programs that do not necessarily have any academic orientation and both of these characteristics are at odds with the design of Project Excel activities.

> ...and so the distinguishing factors are becoming more and more evident as time goes by, that this [Project Excel] is an enrichment after-school program that is science- and literacy-based. And it's not a drop-in program or daycare center situation.

Project Excel as Viewed by the Staff of the Teaching Academy

From the point of view of the Teaching Academy, Project Excel constitutes much of the sophomore year of the Academy's four-year program that introduces students to teaching as a potential career. In keeping with this role, the staff is primarily concerned with teen leaders and their experiences in high school and elementary schools. In Year III, the staff continued to refine the cross-age tutoring arrangements.

Teens continued to work in teams but this year, teens selected one partner and then the Academy staff put pairs together to make four-person teams. The teens selected co-leaders within each team. Another change resulted in teams being assigned to one elementary class for the duration of the field experience. This change, in addition to implementing the program in only two elementary schools, reduced transportation and other logistical problems somewhat from previous years. Because of staffing changes at Independence High School, there was some change in the day to day operation of the cross-age tutoring program. One staff member took major responsibility for the field experiences:

> My role changed a lot in terms of work in the [elementary] classrooms, allowing time for preparation, debriefing, reminding the kids, and also going out to observe them. That increased. In years past, Jane and I would sort of take turns going to observe, and we saw that time when the kids went off to teach Excel as time that Jane and I could also meet on campus, and coordinate our lessons and planned units. But this year it was all my responsibility to supervise on the bus, to make sure that everyone got on and off, and I would observe at one school one day and the other school the following week.

Teams typically made 14 trips to the elementary schools. When asked about the impact of the cross-age teaching experiences on teens, increased personal responsibility was mentioned as a key element:

> I see them changing their focus from being very self centered...to start to extend themselves more and be interested in the welfare and well-being of the smaller kids. They have to be more responsible. They have to be more adult. And whether they were all of those things at all times is debatable. In fact, sometimes it just didn't happen. Some kids were extremely good, and other ones really fell short and even opted to be absent on the days when they were supposed to teach. But generally they became more adult, more serious, and more responsible as time went on.

The Academy staff reported that high school students did not participate in community service projects through Project Excel with the major exception of work done through the 4-H Council at Independence High. The Academy staff was not involved in these activities on a day-to-day basis. Academy staff also noted the decrease in participation in the volunteer scientist program. In Year III, this element of the program had only a handful of volunteers. While the people who did participate were highly valued and went to great lengths to fulfill their roles as mentors, this part of the program was described as being in decline. Project Excel had used volunteer scientists as mentors mostly in the cross-age tutoring component of the program. By Year III, participation of volunteer scientists dropped off considerably and both the staff of the Teaching Academy (and the university-based staff) expressed disappointment at this lost opportunity.

The staff of the Teaching Academy were anticipating changes in the sophomore program for subsequent years. The staff suggested continued "slimming down" of the cross-age tutoring program, possible use of different curricula (for example "Green Circle"), having teens spend more time on preparation and less time in actual teaching, and possibly including activities for teens that would increase their communication and cooperative work skills. The Academy staff suggested increasing the amount of training that teens receive including more coaching and feedback on the enacted lessons. The Academy staff viewed the elements of Project Excel beyond the cross-age teaching component as opportunities that are available to teens but it was entirely up to individual teens whether or not they participated in these additional opportunities.

Project Excel as Viewed by Participating Elementary School Teachers and Administrators

Both the in-school cross-age teaching component and the expanded after-school component of Project Excel continued to be viewed positively by school administrators. The cross-age teaching component was well know by Year III and when asked about it, one building principal commented:

> That's great, that's great. I mean they're [the teens] right where it counts— with the kids. I've seen it in operation and it's great.

But the after-school program and the programs involving parents were seen as an integrated and compelling part of the overall program. One principal praised the after-school program because of its educational value:

> Well, they do science, they do cooperative learning, teaming on various projects, and a lot of self esteem goes into it. It's not just, you know, play time, where they just come in and kick back. It's very instructional and at the same time it's recreational.

Project Excel was named as the source of a number of parent participation activities around the school. In the words of one of the principals:

◆ ...we have "Family Math" that goes on for four Wednesdays towards the end of the year...we invite our parents and our children, students, and put them together as a unit and we work with them on some math skills. Excel has provided training, has provided materials, and also has provided the volunteers that help with the program.

◆ Now insofar as cultural kinds of things tied into Excel: We've organized some folk dancing...so two days a week they [parents] conduct classes to teach them how to dance. And we already had a performance last Friday. So this is part of our parent project. This is done by parents, it's not the staff involved in it. It's the parent running the program. So it's folkloric dancing and there's also a choir. And, along that line also, we have another form of recreation. We have one of our parents teaching the students how to play ball—softball. He gets them together on Friday and weekends—he doesn't work Friday's—and he gets them together and goes over some skills. Basically he's looking at team work, team membership, that kind of thing.

◆ It's kind of a whole...kind of an offspring of the Excel kind of involvement, but, like I said, this is not staff, this is parents running the project.

Focus groups among elementary school teachers revealed some common themes and some differences. When asked about the cross-age teaching that was taking place in their classrooms, there were generally positive responses:

◆ I think it was more streamlined this year. I think the lessons were a little more directed than they were last year. I like the fact that the kids [teens] stayed the whole time—I got to kind of bond with them and get to know them real well. But I thought it was cleaned up from the year before...there was one objective, it was focused right, it was easier for the kids to follow along with what they were doing.

◆I must say that the program was excellent, it was fun, educational. They showed a very positive attitude and I just wish for the program to be repeated next year if that's possible.

◆ I felt I had a good team. They were strong. They were generally very prepared when they came. They worked pretty well with the students.

◆They worked as a team. They had a very positive attitude. They were flexible. They welcomed comments that I made about things that they were talking about. And they also represented different ethnic groups. I saw a big variety in the teams that came into the classroom. And many of them spoke Spanish so they spoke to some of my students in their first language.

◆ And the kids [elementary students] admired them [the teens]...I like [the program] mostly...I like the science and hands on, but mostly in the two years, I think, I like it because it's a role model. These kids are high school kids coming in to our fifth grade class presenting an academic caring role model of a teenager, you know, and they're talking about what they're going to do when they go to college, and talking about what they're studying. It was a good role model for the kids—to hear that the teens are concerned about their grades and their progress and things like that. And some of that comes out, and I think that's really beneficial to our kids—they start to think about their futures.

While the main themes were definitely positive, there were some negative comments as well:

◆ We did larger blocks of time, and there'd be a little period when they didn't come. [There was a] loss of consistency—I didn't think it was as strong as it was the first year.

◆ There wasn't that much enthusiasm this year.

◆ If I do have a criticism, I think the scheduling would be my main, major concern.

◆ Generally speaking, they were strong, but I had one or two that kind of said nothing, did nothing.

When asked about the after-school component of the program, there was a range of responses from confusion about the program at one school to dissatisfaction and criticism at the other. While teachers mentioned that there was the Excel after-school program, it was sometimes confused with the recreational programs (not part of Excel) that was operating at the same schools. There was also an awareness but lack of clear understanding about some of the activities sponsored by parent committees in the schools. Teachers at both schools made comments like:

◆ I'm not so familiar with the after school program.

◆ I don't know about that program.

◆ But as a teaching staff, we were unaware of what was going on. We weren't involved in that part.

◆ I really never was real clear on what was going on and who was in charge. There seemed to be a lot of running around on the playground after school that I wasn't sure was really well supervised. And things that normally as teachers we would say, "don't do," were carried over into recess—like playing with balls on the walls and things.

Teachers suggested that part of the confusion could be cleared up by appropriate communication. In spite of these considerations, teachers were clearly in favor of continuing to work with Project Excel. They view the overall contribution to the children as "time and effort well spent." Teachers praised Project Excel for making it possible for students to participate in several field trips.

Project Excel from the Perspective of the Volunteer Scientists

In Year III, the volunteer scientist program declined to its lowest point since the inception of Project Excel. While two mainstays of the original volunteer pool spent long hours visiting schools and coaching teens in person and by telephone, the project was unable to maintain the participation of the previous year's volunteers or identify new recruits. There were several contributing factors to this situation. First, the science and technology industries in the area were in a general period of downsizing and reorganization. This economic reality placed more pressure on companies and individuals for higher productivity and reduced time available to work on volunteer projects. This factor was identified by the remaining volunteers but there was another condition that also mitigated against continued development of the volunteer scientist component of the program. Project Excel had difficulty in finding or developing an appropriate role for the scientists. Ostensibly they were to mentor the teen leaders as part of the cross-age teaching thrust. However, most scientists were ill-suited for this role and scientist volunteers who tried to use their scientific content knowledge in the context of the program found that that was only marginally useful.

The idea of volunteer scientists as mentors of teens was part of the original design of 4-H SERIES (the science content, pedagogy, and training model that was originally adopted by Project Excel) but although mentor training was partially developed, the role was not defined clearly enough to attract and engage more than a few volunteers. As the 4-H staff (who were primarily involved in the volunteer scientist arrangements) directed increasing amounts of their time and energy toward community development (where volunteer scientists were not a prominent part of the design) and the Teaching Academy was unable to expand the volunteer program as part of its already heavy commitments, participation of scientists and engineers was soon on the decline. By the end of Year III, there was little energy remaining in this area of the project (with the exception of the highly dedicated pair of volunteers mentioned earlier).

Project Excel and the After-School Program

The after-school program became a larger and more important part of the Project with each passing year. By Year III, each of the two elementary schools had a pair of staff members for implementation of daily services for children. Both programs made great progress and it is against this generally positive background that the after-school program should be viewed. There is no doubt that the children liked the program and continued to attend throughout the year. Parents voiced strong support for the program and were anxious to have it continue. The university-based staff have invested heavily in this component of the overall program and after one year were still enthusiastic about its accomplishments and potential. Each program had appropriate attendance levels, had expanded their activities with teen and parent volunteers, and had begun to establish links with the school administration. In terms of the latter, there was a setback at one school when school keys were "lost" and the school restricted privileges for the after-school program somewhat.

While the after school program was generally successful, there were areas that required redirection. From interviews and focus groups on the after-school program, there was some ambiguity expressed about the identity of the program. For example, there were indications that the after-school program was hard to differentiate from the other after-school activities that were available to students in the schools. This was especially true of the drop-in program run by the City of San Jose. While the staff of the Excel Program was clear about differences in the programs, some teachers in the schools, some parents, and perhaps the children themselves had difficulty separating the Project Excel after-school program from other after-school activities. The staff had concerns that are typical of the start-up phase of any new program element. They requested more training, help with logistics, better communication with school site people, more appropriate space for the program, and more time to prepare for activities with the elementary students. There was also some concern in terms of attracting staff for the program that could be expected to perform the broad array of tasks necessary to run the program.

Parents at both schools were strong supporters of the after-school program. In both cases, parents appeared to have detailed knowledge about what their children were doing in the program and were overwhelmingly in support of the additional opportunities for their children to learn in a supervised environment. At one of the sites, a parent 4-H council had been organized and had initiated a number of additional activities for students after school.

Project Excel as Viewed by Teen Leaders

As noted earlier, Project Excel constitutes the sophomore year of the Teaching Academy's program. Teens participate in the in-school portion of Project Excel but they also may participate in a variety of other activities that arise through Project Excel. This section is divided into two parts: (a) responses of sophomores who participated in Project Excel during its third year (based on four focus groups with 30 sophomores, portfolio reviews, and examination of teaching logs describing teen reflections on their weekly teaching experiences); and (b) comparison of perceptions of teaching by third year participants with their views while they were pre-Excel ninth graders (based on a survey of 42 teens in May 1994 and 47 in May 1995).

The primary participation role for teens is in the cross-age teaching program. By enrolling in the Teaching Academy, teens have chosen to explore teaching as a potential career. The entire academy cohort is trained in 4-H SERIES curricula and goes approximately once a week to an elementary school to work with elementary school students. The organizational arrangements for cross-age tutoring varied somewhat from year to year. By Year III, teams of teens were composed of two self selected pairs of students. While students selected partners to make up pairs, the Academy staff then assigned pairs to four-person teams. Co-leaders for each team were chosen by the students. Each team worked in a particular classroom instead of moving from class to class. This arrangement was intended to allow teens to get to know individual students, elementary teachers, and the classroom situations. In Year III, three of the 4-H SERIES curriculum units (snails, chemicals, and recycling) were implemented in two elementary schools.

Teens Views of Cross-Age Teaching in Project Excel

Based on analyses of focus groups and portfolios, a clear pattern emerged in teen's views of teaching. Teens found the experience of cross-age teaching highly challenging. In spite of the level of difficulty (and perhaps because of it), most (but by no means all) teens found the experience rewarding.

Teens enjoyed working with younger children and, like teachers everywhere, were moved by the curiosity and learning that they observed in their "students." Teens also got an immediate sense of how hard the task of teaching is and that the responsibilities associated with teaching are not to be taken lightly. In terms of the performance of teaching, teens were surprised by the amount of preteaching work that must be done in order to even have the possibility of a "good" class. This reality was described as sobering and several teens commented that their experiences in the elementary schools gave them new and more compassionate views of their high school teachers.

Teens liked the majority of the 4-H SERIES units and several saw science more positively after their attempts to teach it. Many teens in the Teaching Academy reported being more interested in teaching *per se* than in teaching science. Perhaps the most frequently addressed topic in the focus groups concerned the dynamics of working in teams of peers. Since planning for teaching and teaching itself were designed to be done in groups of four (and sometimes more), teens were thrown into some interesting circumstances that gave them cause to reflect (or not reflect) on leadership in the classroom. Issues of control, power, and communication in the teaching teams were highly engaging (and often problematic) for the teens. In the focus groups, there were more comments on negative than positive aspects of these issues. The fact that one or more group members did not show up for a teaching occasion made issues of participation and leadership particularly real since there were nontrivial consequences. That is, the group members who did show up had to do the best they could regardless of the preplanned division of work. The teamwork issues made a major impression on teens since the cross-age teaching program created many opportunities for teens to identify with groups and perform publicly with competence—two central themes of adolescent development.

Survey of Teens' Views of Teaching (1994-1995)

In May of 1995, 47 members of the sophomore class of the Teaching Academy were surveyed about their perceptions of teaching, working in groups, and their characteristics as learners. The same survey had been used to assess the perceptions of these same students when they were freshmen (in May 1994). The 1994 sample included 42 participants. These two data sets represent teen's views before they participated in Project Excel (but had completed one year in the Teaching Academy) and after their year of cross-age teaching as participants in Project Excel respectively.

Two survey questions concerned working in groups—a topic that was known from interviews and focus groups to be an issue with teens. Surprisingly, teens responded about the same on the two occasions when asked about working in teams. But when asked about specific actions in groupwork, a pattern emerged. After Excel, teens reported that they took less responsibility for getting their groups to meet than was the case before their participation. They also reported doing homework to support team projects less often and reported taking a leadership role "always." In describing themselves as learners, after participation in Excel teens reported: being able to get along with peers more often; participating in class activities more often; and having to be encouraged to do homework less often.

When asked about how they felt about doing science activities, teens reported a higher likelihood of taking more science courses in high school after participation in Project Excel. Teens also reported being more prepared to speak in front of their peers and to lead a whole class of children. After participation in Excel, more teens reported that community service was an important (even undervalued) activity and that community service

Students demonstrate constructions created with toothpicks, straws, and marshmallows.

work increased their interest in school. However, they also reported that such service should not be required of high school students. Mathematics and English continued to be the most frequently stated "favorite subjects" both before and after the Excel experience with "science" listed in third place.

The same 47 teens who responded to the survey in May 1995 also reported on changes in their: ability to work with various categories of people; interests; and knowledge. Teens reported substantial increases in: abilities to work with peers and children and to organize and carry out tasks; interest in and knowledge about teaching; feelings of being appreciated and contributing to others; leadership skills; and knowledge of science. They reported relatively little change in interest in science and in community issues.

Project Excel's Influence on Elementary School Youth

This section explores two general areas: first, the global impact of Project Excel on third, fourth, and fifth grade students in the target elementary schools; and, second, the attitudes of elementary students toward science and scientists. Data are taken primarily from focus groups conducted with parents and students.

Elementary Students' Responses to Project Excel

Students in the focus groups had clear memories of the teens and the 4-H SERIES science lessons conducted by them. Elementary students accurately reported how many teens came to the class, when they came, how often they came, if there were other adults with them and so on. Students reported liking the visits by teens and the 4-H SERIES lessons. In the initial year of Project Excel, there was evidence that elementary student attendance was higher on days when teens were scheduled to visit their classes.

Similarly, parents were highly supportive of Excel and, while they distinguished easily between the cross-age teaching component and the after-school component, they saw both as critical elements in improving the quality of their children's education. Parents reported that their children often came home from school and recounted what they did and what they learned in the hands-on science lessons of the cross-age teaching program. Several parents commented that they valued participating with their children in one or more of the community events developed and promoted by Project Excel.

Attitudes towards Science among Third, Fourth, and Fifth Graders

Project Excel's goals for elementary school students included introducing them to "hands-on," "heads-on," science activities, focusing on science processes as well as science content, relating science content to other curricular areas and analyzing new learning to local community situations. These goals were initially pursued primarily through the cross-age teaching component of the program. As the project unfolded, the community development aspect of the program broadened the goal structure considerably. Among its goals, the program intends to shift affective orientations of students toward science activities and make careers in science a more viable possibility. One element of students' affective orientations is their attitude toward science. That is, when given choices about alternative activities, do they choose science activities; do science activities constitute a source of satisfaction for them; are they enthusiastic about engaging in science activities; do they represent scientists as social contributors; and do they regard scientific knowledge and processes positively?

Students' attitudes to science and scientists were assessed using two independent measurement procedures—a modified version of the Projective Test of Attitude (PTOA) and the Draw-a-Scientist Test (DAST). The PTOA was used as part of the assessment of Project Excel in each of its first three years of operation. The DAST was used in Years II and III. The results described below are taken from Year III and are illustrative of the trends in students' attitudes.

Data Collection Using the DAST

For the DAST, respondents were asked to draw a representation of a scientist on a single page of paper. In Year III the procedure was administered to 16 classes of third, fourth, and fifth graders in three schools. Responses were analyzed for a number of characteristics that have been related to attitudes to science (Mason, Kahle, and Gardner, 1991).

Teen leaders administered the DAST in October 1994 before they began working on 4-H SERIES activities with the children. Students worked on the task in their regular classrooms. Toward the end of the school year (May 1995), students in the same classes made a second drawing. October and May drawings were collected in three schools (school A which did not participate in Project Excel—three classes, one each of grades 3, 4, and 5; school B—one grade 3, one grade 3/4 combination, two grade 4, and two grade 5 classes; and school C—one grade 3, two grade 4, one grade 4/5 combination, and three grade 5 classes). A total of 733 drawings were coded and analyzed. Two types of coding were completed.

First, each drawing was coded for the gender of the scientist represented. Responses were coded as male, female or other. The "other" category included cases where no human or other live entity was depicted, where the gender of the scientist could not be determined, and where both a male and a female depiction were included in the same drawing. The second type of coding was done globally for each drawing. Following Mason, Kahle, and Gardner (1991), drawings were coded as sinister (involving threat, violence, or apparent destruction), eccentric (unusual, weird, or goofy characterization of the scientist), neutral, or other (uncategorizable in the other three categories).

Gender of Scientists Drawn by Project Excel Participants and Non-Participants

For male students, scientists were depicted overwhelmingly as males regardless of participation in Project Excel. For female students participating in Excel, scientists were depicted as female approximately 10 percent more often in spring compared with fall. A similar trend was found for female students who did not participate in Excel though not as consistently (grade 5 actually dropped). Among females, there was a general decrease in the portion of scientists depicted as female with advancing grade level. That is, females in early grades were likely to depict scientists as female but with experience in school and elsewhere, progressively increased the frequency with which they depicted scientists as male. There was some evidence that, although Project Excel did not halt this trend, it appeared to reduce the effect somewhat. This statement must be taken with caution, especially since the numbers of students at individual grade levels in the non-participating classes were quite small.

Number of Sinister or Eccentric Scientists
Drawn by Project Excel Participants and Non-Participants

For both male and female Excel participants, the proportion of scientists depicted as sinister or eccentric increased from fall to spring. This trend was moderately strong for males and weak for females. There was less change and less consistency in direction of change among the non-participants in Excel. The tendency to depict scientists as sinister or eccentric appeared to be increasing slightly with grade level for both males and females. The proportions were consistently much higher for males than for females.

The Projective Test of Attitudes (PTOA)

The PTOA was adapted from the work of Larry Lowery (1966, 1967). The version of the PTOA used here represents a relatively small part of the larger process developed by Lowery. Individual students were shown a card with a line drawing depicting a youth looking at an experimental set up (a balance beam with items on the pans). The student was asked why the depicted youth was looking at the apparatus, what the depicted youth thought, whether the experiment would be successful, what the depicted youth would do next, and how the depicted youth felt about the experiment.

After the student had responded to the "experiment" card, a second card depicting a youth reading a newspaper story about a new science discovery was shown. In this case, the student responded to four questions about the depicted youths' thoughts and actions regarding the story. The questions were similar to the questions used for the previous card.

The cards shown to a given student depicted a youth of the same gender and ethnic background as the student being interviewed. The interviews were done in English or Spanish depending on apparent comfort level or at the request of the student. Each interview took about five minutes. The procedure was identical to that used in Year II of Project Excel (Fisher, 1994) but slightly modified from that used during Year I (Fisher, 1993).

Data Collection Using the PTOA

PTOA interview data were collected in nine classes in three schools (one grade 3, one grade 4, and one grade 5 class in each school). Within each class, six students (three males and three females) were selected for interviews. Two of the schools (six of the classes) were participants in Project Excel. The third school (three of the classes) were selected from a nearby elementary school that did not participate in the project but served children from similar backgrounds.

Each of the students in the sample was interviewed in the fall (October 1994) as the program was just beginning its activities in elementary schools for the year and again in spring (May 1995) after the cross-age tutoring activities were completed for the year. One hundred and eight interviews were recorded and transcribed. Where students had either a fall or a spring interview missing (for a variety of reasons), they were not included in the analysis. Ninety interviews (45 students) from the three schools were analyzed.

Table 1

Mean Attitude toward Science Scores (PTOA)* for Elementary School Students in Project Excel and Comparison Classes by Gender

	Project Excel		Comparison Classes	
	Oct. 94	May 95	Oct. 94	May 95
Females	5.1 (15)	5.3 (15)	3.1 (6)	6.0 (6)
Males	5.1 (16)	6.5 (16)	3.0 (8)	4.7 (8)
Total	5.1 (31)	5.9 (31)	3.0 (14)	5.3 (14)

*A modified version of the Projective Test of Attitude (Lowery. 1966; Fisher, 1993).
Table contains data for 45 students in nine classes. Numbers in brackets indicate number of students in each cell.

Results on the PTOA

When the categories of students' responses were weighted and summed, the scores ranged from -2 to 9 for both females and males (the maximum range possible was -9 to 9). Mean scores on the revised PTOA for Project Excel and comparison classes by gender are presented in Table 1. The number of students in each cell is shown in parentheses. Inspection of Table 1 reveals that students' attitudes to science as measured by the PTOA increased over the October to May period both for participants and non participants in Project Excel. While the gain was higher for the comparison students, Project Excel students began with much higher attitude scores and ended the school year with higher scores than students in comparison classes. When gender groups were examined separately, the same overall pattern held for both boys and girls. Note that for girls, the increase in attitude scores was very high among the comparison group and, although they started out much lower than the Excel girls, their scores were actually higher at the end of the year. For boys, although both Excel and comparison groups increased, Project Excel boys remained much higher at the end of the year. The initial scores for the Excel participants could have been elevated from previous participation in the Project. Note that the numbers of students in the four cells of Table 1 are not large and therefore interpretation must remain tentative.

Using the PTOA (modified) in grade 3 was probably pushing the lower boundary of the test's applicability. At this age level there were more responses like "I don't know" or "no response" compared to the higher grade levels. In several instances, grade 3 students did not appear to understand the word "successful" as used in the interview. These considerations made the data especially suspect at grade 3. In spite of this problem there was some evidence that students in Project Excel develop positive attitudes toward science more quickly and to higher levels that students in comparison classes.

Discussion

Project Excel has been (and is) a remarkable undertaking. The project has a number of strengths, many of which have been described in earlier sections of this chapter. In terms of the project's original goals, hundreds of elementary school students have had increased opportunities to engage in innovative hands-on science activities, there is evidence that the younger participants have more positive attitudes toward science, high school students have received valuable experience in working with younger youths in classroom settings, and new programs for elementary school-age children and their parents have been established in the local community.

Project Excel is entering its fourth year of operation and is expected to continue to change and improve. This discussion then portrays the initial years of the project and comments on the start-up and early evolution of the program. The project has been generally successful in delivering on its early goals. During the initial years, the project benefited from a strong leadership team that remained reasonably constant from year to year. Since the project involved cooperative agreements among several independent agencies, leadership issues have always been important. As the community development aspect of the project has expanded, these leadership capacities have become even more critical.

Project Excel has introduced a remarkable number of opportunities and programs to elementary school children and their schools as well as to the teen leaders participating in the Teaching Academy. Initially the primary focus of the programs was on cross-age tutoring in science and the leadership opportunities for teens. As the project developed, what began as a secondary focus on community development increased in importance providing a wide range of after-school academic programs and community service activities that engaged both school children and their parents in learning activities. Some of these events were small "family" oriented events and some were large day long programs involving hundreds of participants at a time. As demonstrated earlier in the chapter, Project Excel is greatly appreciated in the community, especially by the children, parents, and local elementary schools.

From this broad community perspective, Project Excel has been successful at building a multilayered overlapping set of programs to support students in and around their local community elementary schools. Accomplishment of these ambitious goals is in large part the result of a clear vision and sustained effort by the Cooperative Extension leaders.

From the narrower perspective of 4-H SERIES, SAY, and science education, the project has made significant impact, but its evolution has taken it on a new course in which science education is still present but no longer central. From the point of view of science education, there are several possible lessons to consider. For example, if science education is to be a primary concern, where is the leadership and support for science education to come from? The cross-age teaching aspect of Project Excel was implemented by the Teaching Academy but no experienced science teacher was available to get involved in the program on a daily basis. This situation was not found in any other SAY sites. The Teaching Academy associated with Project Excel and the students it attracted appeared to be primarily interested in teaching but not necessarily in science teaching. However if these students become elementary school teachers and therefore teach science as part of their classroom teaching responsibilities, they will have had positive early experiences with hands-on science teaching as part of their participation in Project Excel. However, the absence of an experienced science teacher on the Academy's staff probably contributed to reduced emphasis on science teaching and the inquiry pedagogy on which the 4-H SERIES curriculum materials were based. Although the elementary school students responded positively to the teens and to the hands-on activities, as each year went by there was less and less support for the 4-H SERIES curriculum in the project. By the end of the third year, there was discussion of adopting alternative materials.

From the perspective of 4-H SERIES and SAY, there are two other aspects of Project Excel that bear examination. Project Excel had an outstanding cadre of volunteer scientists and engineers. Identifying an effective role or set of roles for this cadre was an ongoing issue. Several individual volunteers were spectacularly successful but their successes were not easily generalized to other volunteers and by the end of Year III this portion of the program was moribund. The notion of a cadre of expert volunteers is common in many educational programs but it is not clear that the details of getting such groups to be productive in their new roles have been adequately worked out. This problem was clearly too much for Project Excel, especially given its ambitious goals, and may also require more resources than 4-H SERIES could afford.

The second aspect of Project Excel that bears consideration from the perspective of 4-H SERIES and SAY concerns community service. Students were engaged in community service projects (for example, tree planting and bread making events) but there were very few examples of community service projects that extended an aspect of the 4-H SERIES curriculum as intended by the curriculum designers. In this particular case, several conditions may have influenced the phenomenon. For example, the teachers and teens participating in the program may have been more interested in the in-school portion of the work and less interested in out-of-school and after-school segments. Community development activities undertaken in the after-school programs may have more or less replaced community service projects that followed the 4-H SERIES design. However, one suspects that this phenomenon is not peculiar to Project Excel but is typical of many school-based programs that resist extending activities beyond the school-yard. Why is this so and what could 4-H SERIES do to encourage more participation in community service activities?

In summary, Project Excel continues to develop into a multifaceted community development project for support of children and their educational and social development. While the role of 4-H SERIES appears to be diminishing in the overall project, it was the primary catalyst during Excel's early development.

Note

1. See chapter 1 for a brief description of the SAY project.

References

Chambers, D.W. (1983). Stereotypic images of the scientist: The draw-a-scientist test. *Science Education, 67,* 255-265.

Fisher, C. (1995). *Project Excel: An assessment of Year III* (September 1994-May 1995). Ann Arbor, MI: Learning Designs.

Fisher, C. (1996). *Attitude toward science among elementary school students in Project Excel* (an Addendum to Project Excel: An assessment of Year III), (September 1994-June 1995). Ann Arbor, MI: Learning Designs.

Fisher, C. (1994). *Project Excel: An assessment of Year II* (September 1993-May 1994). Boulder, CO: Learning Designs.

Fisher, C. (1993). *Project Excel: An assessment of Year I* (September 1992-May 1993). Boulder, CO: Learning Designs.

Jorgensen, E. (November, 1992). *Project Excel Quarterly Report.* University of California Cooperative Extension, Santa Clara County.

Lowery, Lawrence F. (1966). Development of an attitude measuring instrument for science education. *School Science and Mathematics, 66,* 494-502.

Lowery, Lawrence F. (1967). An experimental investigation into the attitudes of fifth grade students toward science. *School Science and Mathematics, 67,* 569-579.

Mason, C., Kahle J., and Gardner, A. (1991). Draw-a-scientist test: Future implications. *School Science and Mathematics, 91*(5), 193-198.

The walls of the science classroom are expanded at the Biosphere near Tucson, Arizona, as 60 teens from across North America come together to share community action projects and learn more about the environment during the 1995 North American Youth Summit.

Students from the Pilgrim Park Child Care Center in Marin City, California, learn about chemical reactions by combining baking soda and vinegar.

Chapter 7

How Informal Science Education Programs Add Capacity to the System

By Richard Ponzio

Introduction

National Education Goal Four—"By the year 2000, U.S. students will be first in the world in mathematics and science"—is certainly one of the most challenging of all the National Education Goals. One need only pick up the daily newspaper to appreciate our society's demand for technological and scientific literacy. With increasing frequency, voters are asked to pass judgment on issues such as offshore oil drilling, the fate of endangered species, and the commercial uses of genetic engineering. Consumers choose, on a daily basis, among products that vary widely in the energy and resource costs used to make, package, and deliver them, as well as their costs in dollars. Employers and employees alike are faced with decisions regarding environmental sensitivity in the workplace. How would increased scientific literacy affect the way these choices and decisions are made? What role does education, in the broad sense, play in increasing day-to-day science literacy of Americans? And, how do we, as an educational community, raise levels of scientific literacy among all segments of American youth?

Science Education Reform

To address this increasingly important demand for science literacy, educators have advocated widespread reform in elementary and secondary schools. While calls for reform in science education are not new, the current wave can be distinguished from earlier reforms in several ways. Reforms in the 1970s and 1980s tended to focus on increasing the amount of time students spent on science in their schooling and on improving access to science education and science careers for female and minority students. The American Association for the Advancement of Science concluded that:

> The present science textbooks and methods of instruction, far from helping, often impede progress toward science literacy. They emphasize the learning of answers more than the exploration of questions, memory at the expense of critical thought, bits and pieces of information instead of understandings in context, recitation over argument, reading in lieu of doing. They fail to encourage students to work together, to share ideas and information freely with each other, or to use modern instruments to extend their intellectual capabilities. The present curricula in science and mathematics are overstuffed and undernourished. (American Association for the Advancement of Science, 1989)

Recent attention has been directed to reforming both the content and pedagogy of school science and maintaining the emphasis on improving access for all Americans to science education and science-based careers. Science education proposes to emphasize the use of scientific thinking processes, the reorganization of science content, and increased attention to presenting applications of the scientific principles being learned to social issues in the homes and communities of school-age students. These shifts have fostered new interest in "hands-on" and "heads-on" approaches to science teaching and learning.

As we approach the 21st century, there is an increasing need for an educational "mosaic," one that is inclusive of the myriad science learning opportunities afforded in the community—a mosaic that is made up of inter-related, reinforcing parts. Certainly schools are one piece, but there are others that are currently underutilized, including after-school child care, youth groups sponsored by community agencies, and programs offered by boys and girls clubs, the 4-H Youth Development program, and institutions such as museums, zoos, etc. There is also opportunity for educational pieces for the mosaic to be offered by parents; sometimes through home schooling, other times as volunteers in programs such as Campfire, 4-H, and scouting.

Linking Learning and Service

Robert Reich (1983, 1991) argues that America is no longer dominated by a production-line economy but is rapidly moving toward a dynamic, entrepreneurial, global economy, and that our schools should provide experiences for learners that are dynamic and entrepreneurial by design. In part, this notion suggests that schooling should include more activities that allow students to work cooperatively on the heuristics of problem finding, problem framing, and problem solving.

Additional impetus for entrepreneurial applications of learning to real-world problem solving can be found in the rekindled interest in service learning, which is defined as the blending of both service and learning in such a way that each is enriched by the other. The support of the federal government for this trend toward the blending of learning and service can be seen in the Congressional reauthorization of the National and Community Service Trust Act of 1993, designed to place up to 24,000 participants in national service assignments that include service learning components.

Limitations of Traditional School Science

Major research efforts on schools and schooling document the inability of traditional school structures to provide students with even minimal opportunities for learning from hands-on community-based science experiences (Goodlad, 1984; AAAS, 1989). In part, this lack of experiential learning in science is based on the way schools are organized and what forms of student learning are assessed. Ellen C. Lagemann, speaking of educational research traditions claims that "one cannot understand the history of education in the United States during the 20th century unless one realizes that Edward L. Thorndike won, and John Dewey lost" (1989).

Each of these factors—*i.e.,* changes in the content and pedagogy of science education, stronger linkages between learning and service, and dissatisfaction with traditional science education—has contributed to the current wave of reform in classroom science instruction and, inevitably, to calls for reform in training of science teachers. These shifts have fostered new interest in "constructivist" approaches to teaching and learning science, as well as an interest in authentic tasks and authentic assessment.

The Place of Community-Based Learning in the Science Education Mosaic

Recent work by researchers such as Howard Gardner (1983, 1991) and Siegel and Shaughnessy (1994), speak to the issue of students' lack of understanding—the inability of students to take knowledge, skills, and other apparent attainments and apply them successfully in new situations. The literature on multiple intelligences suggests multiple paths toward a goal, including involvement in projects—either projects assigned to them or projects they have helped design—and multiple outcomes as expressions of student understanding, including the use of student portfolios. An educational mosaic metaphor suggests we invent a design for education that includes a broad view of education that, by necessity, takes a look at all the educational resources available within, and beyond, the classroom walls. In the remainder of this piece we will examine the contributions of community-based education programs to the science education mosaic.

Unique Opportunities in Community-Based Programs

Community-based science programs provide a way of engaging in literacy development in a broader context than that available in traditional schooling. They are beyond the school fences, beyond the school bells, and bridge the gap between school and popular culture. Successful community-based informal science education programs, by necessity, have some things in common: they're **fun**, they have to be **interesting**, and they're usually **easily accessible** and **cheap**.

In any event, a unique aspect of community-based education programs is that both the leaders and the youth participants have the right to "vote with their feet" in terms of participation in the programs. Perhaps one of the most powerful aspects of community-based science education programs is their flexibility. This flexibility manifests itself in terms of time (dates, contact hours, frequency, and duration), topics, cognitive demands, ages of participants, and expected outcomes.

Flexible Time

Although most community-based informal science programs are offered at regularly scheduled times, the opportunity to meet at the convenience of the participants is almost always there. Most community-based programs are sensitive to the time demands of their targeted clientele.

Flexible Topics

Most community-based educational programs are justified and sustained by participants enrolled in the program. This is usually a function of the participants' interest in the topic. Such a user-driven system requires the designer to take into account the interests and ability levels of the intended audience.

Variable Cognitive Demands

Many community-based programs are designed around experiential activities targeted to various age levels of the participants. That is, even within a specific content area such as "Insects and Spiders" there will be a group of hands-on activities targeted to young children that is quite different in its cognitive and social demands than activities prepared for older children or teenage youth. The various hands-on activities allow participants to increase their confidence and competence in using the processes of science, and expand their repertoire of hands-on experiences that lead to heads-on science learning.

Cross-age Instruction

Most community-based science literacy programs are made up of participants of different ages and abilities. Current research emphasizes the value of social interaction for improving learning (Covington, 1992; Slavin, 1983). Community-based programs build in opportunities for youngsters to learn science from each other—older teens interacting with younger learners to solve problems, record data, make inferences, and so on. The modeling that older youngsters provide can often be more effective than the modeling provided by an adult teacher because the age differences are less, and the time for personal, more individualized contact is increased. With less age and status difference, a truly two-way interaction starts faster and generates more enthusiasm.

Variable Outcomes

If one follows a constructivist educational model that views direct experience to be antecedent to learning, the case is made for direct experiences in science. It follows that participants would then work on authentic tasks using their newly acquired skills in new ways. This paradigm fits well with proven pedagogical practices, such as the learning cycle (Guzzetti, 1992; Karplus, 1967; Lawson, 1989), and cooperative learning strategies (Covington, 1992; Slavin, 1983) that have been found to be effective in science instruction.

Community-based science programs also allow participants to apply their learning to a wide variety of home, neighborhood, and community situations in settings such as helping to design and implement recycling programs, raising vegetables in community gardens for senior citizens centers, or helping design family disaster-emergency response plans. These projects encourage youth to solve problems that are grounded in real-world contexts requiring many kinds of complex problem solving skills suggested by advocates of "outcome-based" education (Spady, 1994). Although the diversity of projects and outcomes poses a major challenge to evaluation, it helps keep participants engaged in service learning applications of their science knowledge. The project outcomes also help forge a connection between "school smarts" and "street smarts" by tying science learning to community issues and applications significant to the learners.

Design Characteristics

Putting together the pieces of the educational mosaic involves design considerations. What are the most appropriate roles of each of the components? What can the schools do best? The museums? The community-based organizations such as 4-H clubs, scouts, boys and girls clubs, Girls Inc., Campfire? How do we design a mosaic that includes the stakeholders? Much of the literature to inform us is there, from the works of Dewey to Jean Piaget, and the more recent cognitive scientists and brain researchers. Many of the instructional models are there, from Socrates to current research on effective instruction and recent work on problem solving by Gardner and David Perkins.

Designing an educational mosaic for science education includes complementary pieces that represent a common goal or vision. It might be described as many paths to the goal of science literacy. Most programs in science education have some common features, regardless of whether they are based in schools, museums, or theme parks. Although the venue may vary, each program can reinforce the development of science literacy through application of principles of learning, scientific investigation, and participant involvement that are common to each:

◆ Use of scientific thinking processes (observing, communicating, comparing, organizing, relating/experimenting, inferring, and applying) found in virtually all school science programs.

◆ Use of the Learning Cycle (exploration, concept introduction, and concept application) developed by Karplus and Lawson, and adapted by Renner (1988) and other authors. Recent work by Guzzetti and others has shown its effectiveness in instructional design including improved student understanding of science textbooks.

◆ Use of activities that are participatory (hands-on when appropriate) and inquiry-based with opportunities for participant reflection, and include questioning strategies that engage participants in making sense of what was observed, and in constructing a mental model or theory.

◆ Use of activities that are designed so that participants engage in cooperative learning. Learning tasks, and community service projects are also structured in a way that requires the group to work together.

◆ Use of authentic assessment opportunities such as the construction of a self-managed portfolio that represents the participant's work. Portfolios often include goal statements, data, and artifacts from the community service projects. The portfolio might also include letters from parents, newspaper clippings, etc. If there is to be a comparison of portfolios, one might want to standardize some features such as a table of contents or a set of required components. Portfolios are assessment tools that bridge the gap between community-based programs and schools, as they become increasingly popular in models such as "outcome based education," and are more widely used for student assessment by schools. Portfolios support a constructivist approach to learning by providing the students with a tangible sense of accomplishment, and allow them to reflect on their actions and outcomes.

In an attempt to provide a concrete example of the educational characteristics listed above being translated into operational practice we offer some hard won examples from the national 4-H SERIES project. The 4-H SERIES project continues to be operational well after the NSF support has expired, and new curricula for the program continue to be developed by state 4-H programs and other agencies.

Theoretical Framework for Curriculum and Activities in 4-H SERIES

The instructional activities in the 4-H SERIES curricula emphasize use of the scientific thinking processes included in virtually all school science texts used in the United States. These processes—observing, communicating, comparing, organizing, relating, inferring, and applying—are used by participants in each of the activities. The teen leaders thus have early experiences in using scientific thinking processes in the 4-H SERIES inquiry-based activities, and in teaching the processes. Each of the curricula is built on a model of instructional design known as the learning cycle (Karplus and Thier, 1967; Karplus, Lawson, Wollman, Appel, Bernoff, Howe, Rusch, and Sullivan, 1980; Renner, and Marek, 1988; Lawson, Abraham, and Renner, 1989). This model, widely used in contemporary science education, has shown consistent learning gains in a variety of educational settings (Guzzetti, Snyder, and Glass, 1992). The learning cycle instructional model suggests three distinct segments for an activity within a lesson.

In the general formulation of the model, each activity begins with an exploration segment during which students manipulate the materials, encounter some interesting phenomena, and perhaps attempt to understand the phenomena by changing something in the environment to see possible effects on the phenomenon itself. In the second segment, the students and leader or teacher engage in a discussion of the events that occurred in the first segment. During this segment, one or more concepts relevant to the phenomenon are introduced and related to the students' prior knowledge. Once key concepts are developed and articulated through discussion, the third segment involves having students apply the concept or concepts in some personally meaningful context. This general structure, exploration followed by concept introduction followed by concept application, is the primary template for the 4-H SERIES curricula. In the project curriculum materials, the first two components of the learning cycle are embedded in the teen-led learning activities, and the third component is developed in the community service activities. The learning cycle structure is explicitly taught to the teen leaders as part of the training associated with the project.

Community Service Projects

The inclusion of community service projects as an integral part of the 4-H SERIES curricula is a crucial component of the program. The project draws on a collaborative working relationship with states' 4-H Youth Development Programs. 4-H is a National Youth Development Program under the auspices of the U.S. Department of Agriculture and administered through each state's land grant college or university system. The service learning component of the project uses the expertise of local paid academic 4-H staff and volunteer leaders to support youth in improving their neighborhoods and homes through community service projects.

There are at least four reasons for underscoring the importance of this component. First, as mentioned above, community service projects constitute the mechanism for completing the third segment of the learning cycle. This does not imply that learning would not be appropriated by students if there were no community service project (or other mechanism to accomplish the concept application segment of the learning cycle), only that a substantially larger proportion of students are likely to successfully appropriate learning in the context of a community service project than would be the case if no such supporting structure were offered.

Second, the community service project represents a learning activity that is expressly initiated by the learners to relieve a perceived need. So much of traditional education is one way, in the sense that the transactions flow primarily from the teacher to the students. In this admittedly stereotypic view, the teacher is the source of knowledge, talk, and tasks in the classroom while students are passive receivers and compliant doers. In contrast, contemporary teaching and learning frameworks are based on a dialogic process, a two-way process, where there is the presence of, if not equity between, both teacher and student voices. Community service projects provide one context within which the student voice is given more expression than is usually the case.

Third, community service projects afford an opportunity for elementary and high school students to make an authentic contribution to their communities. There seem to be fewer avenues for young people to make legitimate contributions to communities than there were even a few decades ago. The extension of formal schooling to what is now essentially a 16-year process may exacerbate the situation. Community service projects, like service learning generally, create a context for students to identify an issue and do something about it.

Fourth, community service projects illustrate how scientific knowledge and processes can be used for social purposes. In the course of inventing and carrying out projects in their communities, students negotiate, make agreements, and depend on one another and members of the community at large.

This action orientation encourages the development of leadership and civic responsibility. While the rationales for community service projects are appealing, it is difficult to describe the impact of participating in them in simple terms. To indicate the kinds of projects that teen leaders and nine to 13 years olds generate, here is an example from a teen-led 4-H SERIES project in California:

> In Santa Barbara County, two groups of youth cooperated in a beach debris comparison project as a culmination of work on the "Recycle, Reuse, Reduce" curriculum. First, group members went to Santa Cruz Island to visit a beach that, on an annual basis, had very few visitors. The group, in one day's effort, collected 600 pounds of debris over a one-mile strip of the beach. On analysis, the debris varied from old tires to fish traps to propane tanks, but the most prevalent items were plastic and Styrofoam. Using the data from this island beach as a point of comparison, each group subsequently took responsibility for monitoring debris on separate onshore beaches near their homes. Each group cleaned a designated one-mile strip of beach and recorded the results once a month. The debris was disposed of in a responsible manner, most of it recycled, and data on the local beaches was reported to county authorities.

Other examples of projects include the distribution of earthquake preparedness information and the strapping of water heaters in earthquake-prone communities, and production of clothing for homeless people.

Interdependent Collaboration with other Agencies and Organizations

4-H SERIES is collaborating with a variety of professional scientific and engineering organizations such as the Society of Women Engineers and the Society of Hispanic Engineers as well as agencies such as the American Red Cross. The collaboration is based on common goals such as the recruitment of youth from groups traditionally underrepresented in the sciences and the opportunity for teens to participate in community service projects and disaster preparedness planning. 4-H SERIES provides a venue for professionals to "coach" teens who are themselves coaching younger children in models of scientific investigation. 4-H SERIES has also collaborated with ASTC's YouthALIVE! program by providing teen and adult training sessions and project activities and materials for YouthALIVE! participants to present in children's' museum settings.

In Summary

An ancient African proverb reminds us, "It takes an entire village to educate a child." The point is that there are multiple contributions to be made by the inclusion of a variety of community resources available to educate our children. If we view the goal of improving science education as an educational mosaic, made up of multiple pieces, each with its unique contribution, then each of our community-based programs can be viewed as an integral part of the bigger picture. Schools, museums, child-care, and community-based programs each have a unique venue and opportunity to reach children and to improve the quantity and quality of the science literacy experiences available to youth. An educational mosaic celebrates diversity of participants, programmatic content, delivery system, and outcome.

References

American Association for the Advancement of Science (1989). *Project 2061: Science for all Americans.* Washington, DC: AAAS Publications.

Covington, M. (1992). *Making the grade: A self-worth perspective on motivation and school reform.* New York: Cambridge University Press.

Gardner, H. (1983). *Frames of mind: The theory of multiple intelligences.* New York: Basic Books.

Gardner, H. (1991). *The unschooled mind: How children think and how schools should teach.* New York: Basic Books.

Goodlad, J.I. (1984). *A place called school.* New York: McGraw-Hill.

Guzzetti, B., Snyder, T., and Glass, G. (1992). "Promoting conceptual change in science: Can texts be used effectively?" *Journal of Reading,* 35, 542-649.

Lagemann, E.C. (1989). "The plural worlds of educational research." *History of Education Quarterly,* 29, 185-214.

Karplus, R., Lawson, A., Wollman, W., Appel, M., Bernoff, R., Howe, A., Rusch, J., and Sullivan, F. (1980). *Science teaching and the development of reasoning.* (4th Printing). Berkeley, CA: The Regents of the University of California.

Karplus, R., and Thier, H. (1967). *A new way to look at elementary school science.* Chicago, IL: Rand McNally.

Lawson, A., Abraham, M., and Renner, J. (1989). *A theory of instruction: Using the learning cycle to teach science concepts and thinking skills.* Monograph, Number One, National Association for Research in Science Teaching.

Reich, R.B. (1983). *The next American frontier: A provocative program for economic renewal.* New York: Penguin Books.

Reich, R.B. (1991). *The work of nations: Preparing ourselves for 21st century capitalism.* New York: Knopf.

Renner, J., and Marek, E. (1988). *The learning cycle and elementary school science teaching.* Portsmouth, NH: Heinemann.

Siegel, J., and Shaughnessy, M.F. (1994) Educating for understanding: An interview with Howard Gardner. *Phi Delta Kappan,* 75 (7), 563-566.

Slavin, R. (1983) *Cooperative Learning.* New York: Longman.

Spady, W.G. (1994) Choosing outcomes of significance. *Educational Leadership.* 51 (6), 18-22.

Notes

Notes